Leadership in Energy and Environmental Design

LEED® Homes Practice Exam

David M. Hubka, DE, LEED AP, and
Vessela Valtcheva-McGee, LEED AP

NEW ENGLAND INSTITUTE OF TECHNOLOGY
LIBRARY

Professional Publications, Inc. • Belmont, California

Benefit by Registering this Book with PPI

- Get book updates and corrections
- Hear the latest exam news
- Obtain exclusive exam tips and strategies
- Receive special discounts

Register your book at **www.ppi2pass.com/register**.

Report Errors and View Corrections for this Book

PPI is grateful to every reader who notifies us of a possible error. Your feedback allows us to improve the quality and accuracy of our products. You can report errata and view corrections at **www.ppi2pass.com/errata**.

PPI is not affiliated with the U.S. Green Building Council (USGBC®) or the Green Building Certification Institute (GBCI). PPI does not administer the Leadership in Energy and Environmental Design (LEED®) credentialing program or the LEED Green Building Rating System. LEED and USGBC are registered trademarks of USGBC. PPI does not claim any endorsement or recommendation of its products or services by USGBC or GBCI.

Energy Star® (ENERGY STAR®) is a registered trademark of the U.S. Environmental Protection Agency (EPA).

LEED HOMES PRACTICE EXAM

Current printing of this edition: 1

Printing History

edition number	printing number	update
1	1	New book.

Copyright © 2009 by Professional Publications, Inc. (PPI). All rights reserved. No part of this publication may be reproduced, stored in a retrieval system, or transmitted, in any form or by any means, electronic, mechanical, photocopying, recording, or otherwise, without the prior written permission of the publisher.

Printed in the United States of America

PPI
1250 Fifth Avenue, Belmont, CA 94002
(650) 593-9119
www.ppi2pass.com

ISBN: 978-1-59126-183-4

Table of Contents

Preface and Acknowledgments v

Introduction vii
    About the LEED Credentialing Program vii
    About the LEED AP Homes Exam vii
    Taking the LEED Credentialing Exams ix
    How to Use This Book x

References xi
    Primary References for Exam Part 1: LEED Green Associate xi
    Secondary References for Exam Part 1: LEED Green Associate xi
    References for Exam Part 2: LEED AP Homes Specialty xii

Practice Exam Part One 1

Practice Exam Part Two 23

Practice Exam Part One Answers 43

Practice Exam Part Two Answers 63

Preface and Acknowledgments

This book, which is written to help you study for and pass the LEED AP Homes credentialing exam, is a product of our passion for green building. Early in our respective careers in architecture and mechanical systems design, we recognized that many contractors, engineers, product vendors, and architects desired a fundamental understanding of LEED. Our intimate familiarity with the LEED rating system and the particularities of LEED for Homes contributed to our ability to create this book—the best available simulation of the actual LEED AP Homes credentialing exam. Though the questions in this book are similar to those on the actual exam, they are unique to us, and are the result of our experiences.

We wish to acknowledge those who have helped us create this book. David's thanks go to the incredibly talented professionals working at Total Mechanical. Their combined experience across all mechanical trades has been an invaluable resource, and their guidance has contributed significantly to his broad knowledge of LEED. He also thanks Mike Hyde, Total Mechanical's Chief HVAC Engineer, who continually provides forward-thinking solutions to LEED projects. David also is deeply appreciative of the Mechanical Service Contractors of America (MSCA) for hiring him to provide his first LEED exam prep training, and of Barb Dolim, MSCA Executive Director, for giving direction as he developed the training. He also thanks all of his LEED seminar attendees, who continually contribute to his understanding and appreciation of the rewards and challenges of green building. Finally, he would like to thank his wife, Dana, for her support throughout the entire process.

Vessela's thanks go to the folks at Southface Energy Institute for giving her the opportunity to serve as a LEED for Homes provider representative, as well as for all the guidance and education they provided in the process of creating this book. She also thanks Neil Dawson, AIA, and Richard Wissmach, AIA, for their professional encouragement and support. Finally, Vessela thanks Kelly Gearhart, LEED AP, for her broad knowledge of LEED and her consistent support.

Together, we wish to acknowledge those at PPI who helped in the process of creating this book, including director of editorial Sarah Hubbard, director of production Cathy Schrott, project editor Courtnee Crystal, proofreader Sesa Pabalan, and typesetter and cover designer Amy Schwertman.

Finally, your suggestions are important to us. As the LEED exam process evolves, so will this book. If your experiences lead you to different answers than those presented in this book, we'd like to hear from you. We request that you report all errata and suggestions using the errata section on the PPI website at **www.ppi2pass.com/errata**. Corrections will be posted on the PPI website and incorporated into this book when it is reprinted.

Best wishes in all your green building efforts!

David M. Hubka, DE, LEED AP, and Vessela Valtcheva-McGee, LEED AP

Introduction

About the LEED Credentialing Program

The Green Building Certification Institute (GBCI) offers credentialing opportunities to professionals who demonstrate knowledge of Leadership in Energy and Environmental Design (LEED) green building practices. *LEED Homes Practice Exam* prepares you for both parts of the LEED AP Homes exam.

GBCI's LEED credentialing program has three tiers. The first tier corresponds to the LEED Green Associate exam. According to the *LEED Green Associate Candidate Handbook*, this exam confirms that you have the knowledge and skills necessary to understand and support green design, construction, and operations. When you pass the LEED Green Associate exam, you will earn the LEED Green Associate credential.

The second tier, which corresponds to the LEED AP specialty exams, confirms your deeper and more specialized knowledge of green building practices. There are four tracks for the LEED AP exams: LEED AP Homes, LEED AP Operations & Maintenance, LEED AP Building Design & Construction, and LEED AP Interior Design & Construction; and the LEED AP Neighborhood Development is being planned. The LEED AP exams are based on the corresponding LEED reference guide and rating systems and other references. When you pass the LEED Green Associate exam along with any LEED AP specialty exam, you will earn the LEED AP credential.

The third tier, called LEED AP Fellow, will distinguish professionals with an exceptional depth of knowledge, experience, and accomplishments with LEED green building practices. This distinction will be attainable through extensive LEED project experience, not by taking an exam.

For more information about LEED credentialing, visit **www.ppi2pass.com/LEEDhome**.

About the LEED AP Homes Exam

The LEED AP Homes exam contains 200 questions. (So does the practice exam in this book.) Exam Part 1: LEED Green Associate Exam contains 100 questions that test your knowledge of green building practices and principles, as well as your familiarity with LEED requirements, resources, and processes related to both commercial and residential spaces and both new construction and existing building projects. Accordingly, GBCI categorizes the Exam Part 1 questions into the following seven subject areas.

- *Synergistic Opportunities and LEED Application Process* (project requirements; costs; green resources; standards that support LEED credit; credit interactions; Credit Interpretation Requests and rulings that lead to exemplary performance credits; components of LEED Online and project registration; components of LEED Scorecard; components of letter templates; strategies to achieve credit; project boundary; LEED boundary; property boundary; prerequisites and/or minimum program requirements for LEED certification; preliminary rating; multiple certifications for same building; occupancy requirements; USGBC policies; requirements to earn LEED AP credit)
- *Project Site Factors* (community connectivity: transportation and pedestrian access; zoning requirements; development: heat islands)
- *Water Management* (types and quality of water; water management)
- *Project Systems and Energy Impacts* (environmental concerns; green power)
- *Acquisition, Installation, and Management of Project Materials* (recycled materials; regionally harvested and manufactured materials; construction waste management)
- *Stakeholder Involvement in Innovation* (integrated project team criteria; durability planning and management; innovative and regional design)
- *Project Surroundings and Public Outreach* (codes)

Exam Part 2: LEED AP Homes Specialty Exam contains an additional 100 questions that test your knowledge of subject areas unique to green home building design and construction. Accordingly, GBCI categorizes the Exam Part 2 questions into the following seven subject areas.

- *Project Site Factors* (considerations for site selection: land issues, plants, and animals; community connectivity and services; development: building and land; green management; climate conditions)
- *Water Management* (water treatment; stormwater; irrigation demand)
- *Project Systems and Energy Impacts* (energy performance policies; building components; on-site renewable energy; requirements and alternative rating systems for third-party relationships; energy performance measurement; energy trade-offs; energy usage; specialized equipment power needs)
- *Acquisition, Installation, and Management of Project Materials* (building reuse; rapidly renewable materials; acquisition of materials; neutral homes)
- *Improvements to the Indoor Environment* (minimum ventilation requirement; tobacco smoke control; air quality; ventilation effectiveness; indoor air quality: pre-construction, during construction, before occupancy, and during occupancy; low-emitting materials; indoor/outdoor chemical and pollutant control; lighting controls; thermal controls; views; acoustics; residential requirements)
- *Stakeholder Involvement in Innovation* (design workshop/charrette; ways to earn credit; education of homeowner or tenant; education of building manager)
- *Project Surroundings and Public Outreach* (preferred location; infrastructure; information of available community resources; site selection in collaboration with developer; zoning requirements; government planning agencies; planning terminology; land

development phases; public-private partnership; development footprint reduction methods; reduced parking methods; transit-oriented development; pedestrian oriented streetscape design; ADA/Universal access; streetscape planning)

Taking the LEED Credentialing Exams

To apply for a LEED credentialing exam, you must agree to the Disciplinary and Exam Appeals Policy and credential maintenance requirements and submit to an application audit. To be eligible to take the LEED Green Associate exam, one of the following must be true.

- Your line of work is in a sustainable field.
- You have documented experience supporting a LEED-registered project.
- You have attended an education program that addresses green building principles.

To be eligible to take a LEED AP exam, you must have documented experience with a LEED-registered project within the three years prior to your application submittal.

The LEED credentialing exams are administered by computer at Prometric test sites. Prometric is a third-party testing agency with over 250 testing locations in the United States and hundreds of centers globally. To schedule an exam, you must first apply at www.gbci.org to receive an eligibility ID number. Then, you must go to the Prometric website at www.prometric.com/gbci to schedule and pay for the exam. If you need to reschedule or cancel your exam, you must do so directly through Prometric.

The LEED credentialing exam questions are multiple choice with four or more answer options for each question. If more than one option must be selected to correctly answer a question, the question stem will indicate how many options you must choose. Each 100-question exam lasts two hours, giving you a bit more than one and a half minutes per question. The bulk of the questions are non-numerical. Calculators are provided, but only basic math is needed to correctly solve any quantitative questions. No reference materials or other supplies may be brought into the exam room, though a pencil and scratch paper will be provided by the testing center. (References are not provided.) The only thing you need to bring with you on exam day is your identification.

Your testing experience begins with an optional brief tutorial to introduce you to the testing computer's functions. When you've finished the tutorial, questions and answer options are shown on a computer screen, and the computer keeps track of which options you choose. Because points are not deducted for incorrectly answered questions, you should mark an answer to every question. For answers you are unsure of, make your best guess and flag the question for later review. If you decide on a different answer later, you can change it, but if you run out of time before getting to all your flagged questions, you still will have given a response to each one. Be sure to mark the correct number of options for each question. There is no partial credit for incomplete answers (or for selecting only some of the correct options).

If you are taking both the first tier (LEED Green Associate) and the second tier (LEED AP) exams on the same day, at the end of your first session the computer will ask you if you are ready to take the second tier. You can take a short break of about 3 minutes at this time. The second tier's two hours will begin when you click "yes" to indicate that you are ready.

To ensure that the chances of passing remain constant regardless of the difficulty of the specific questions administered on any given exam, GBCI converts the raw exam score to a scaled score, with the total number of points set at 200 and a minimum passing score of 170. In this way, you are not penalized if the exam taken is more difficult than another exam. Instead, in such a case, fewer questions must be answered correctly to achieve a passing score. Your scaled score (or scores, if you are taking both tiers on the same day) is reported on the screen upon completing the exam. A brief optional exit survey completes the exam experience.

When you pass the LEED Green Associate exam, a LEED Green Associate certificate will be sent to you in the mail. If you take and pass both exams, a LEED AP certificate will be sent to you in the mail. If you take both exams but pass only the LEED AP exam, you will need to register again and retake and pass the LEED Green Associate exam before you receive any LEED credential.

How to Use This Book

There are two ways you can use this book's practice exam. You can determine your areas for further study with an untimed review of the questions and answers. Familiarize yourself with the exam format and content and determine which subjects you are weak in. This book's companion volume, *LEED Prep Homes: What You Really Need to Know to Pass the LEED AP Homes Exam*, will give you a complete, concise review of the subjects covered on the exam. *LEED Homes Flashcards*, also available through PPI, will reinforce your ability to retain and recall what you've studied.

Or, you can use this book to simulate the exam experience, either as a pretest before you begin your study or when you think you are fully prepared. In this case, treat this practice exam as though it were the real thing. Don't look at the questions or answers ahead of time. Put away your study materials and references, set a timer for two hours, and answer as many questions as you can within the time limit. Practice exam-like time management. Fill in the provided bubble sheet with your best guess on every question regardless of your certainty and mark the answers to revisit if time permits. If you finish before the time is up, review your work. If you are unable to finish within the time limit, make a note of where you were after two hours, but continue on to complete the exam. Keep track of your time to see how much faster you will need to work to finish the actual exam within two hours.

After taking the practice exam, check your answers against the answer key. Consider a question correctly answered only if you have selected all of the required options (and no others). Calculate the percentage correct. Though the actual exam score will be scaled, aim for getting at least 70% (70 questions) of the practice exam's questions correct. The fully explained answers are a learning tool. Therefore, in addition to reading the answers to the questions you answered incorrectly, also read the explanations of those you answered correctly. Categorize your incorrect answers by exam subject to help you determine which areas need further study. Use GBCI's list of references and PPI's *LEED Prep Homes* to guide your preparation. Though this practice exam reflects the breadth and depth of the content on the actual exam, use your best judgment when determining the subjects you need to review.

References

LEED Homes Practice Exam is based on the following references, identified by the Green Building Certification Institute (GBCI) in its *LEED AP Homes Candidate Handbook*. Most of these references are available electronically. You can find links to these references on PPI's website, **www.ppi2pass.com/LEEDreferences**.

Primary References for Exam Part 1: LEED Green Associate

Bernheim, Anthony, and William Reed. "Part II: Pre-Design Issues." *Sustainable Building Technical Manual*. Public Technology, Inc. 1996.

Cost of Green Revisited: Reexamining the Feasibility and Cost Impact of Sustainable Design in Light of Increased Market Adoption. Sacramento, CA: Davis Langdon, 2007.

Guidance on Innovation & Design (ID) Credits. Announcement. Washington, DC: U.S. Green Building Council, 2004.

Guidelines for CIR Customers. Announcement. Washington, DC: U.S. Green Building Council, 2007.

LEED for Building Design & Construction Reference Guide. Washington, DC: U.S. Green Building Council, 2009.

LEED Technical and Scientific Advisory Committee. *The Treatment by LEED of the Environmental Impact of HVAC Refrigerants*. Washington, DC: U.S. Green Building Council, 2004.

Secondary References for Exam Part 1: LEED Green Associate

AIA Integrated Project Delivery: A Guide. American Institute of Architects, 2007.

Americans with Disabilities Act: Standards for Accessible Design. 28 CFR Part 36. Washington, DC: Code of Federal Regulations, 1994.

"Codes and Standards." Washington, DC: International Code Council, 2009.

"Construction and Building Inspectors." *Occupational Outlook Handbook*. Washington, DC: Bureau of Labor Statistics, 2009.

Frankel, Mark, and Cathy Turner. *Energy Performance of LEED for New Construction Buildings: Final Report*. Vancouver, WA: New Buildings Institute and U.S. Green Building Council, 2008.

GSA 2003 Facilities Standards. Washington, DC: General Services Administration, 2003.

Guide to Purchasing Green Power: Renewable Electricity, Renewable Energy Certifications, and On-Site Renewable Generation. Washington, DC: Environmental Protection Agency, 2004.

Kareis, Brian. *Review of ANSI/ASHRAE Standard 62.1-2004: Ventilation for Acceptable Indoor Air Quality*. Greensboro, NC: Workplace Group, 2007.

Lee, Kun-Mo, and Haruo Uehara. *Best Practices of ISO 14021: Self-Declared Environmental Claims*. Suwon, Korea: Ajou University, 2003.

LEED Steering Committee. *Foundations of the Leadership in Energy and Environmental Design Environmental Rating System: A Tool for Market Transformation*. Washington, DC: U.S. Green Building Council, 2006.

References for Exam Part 2: LEED AP Homes Specialty

Carlson, Spike. "Making Your House Quieter." The Family Handyman, 2000.

Energy Star Qualified Homes Thermal Bypass Inspection Checklist. Washington, DC: U.S. Environmental Protection Agency, 2008.

Introduction to Indoor Air Quality: About Carbon Monoxide Detectors. Washington, DC: U.S. Environmental Protection Agency, 2009.

LEED for Homes Reference Guide. Washington, DC: U.S. Green Building Council, 2008.

LEED for Homes Rating System. Washington, DC: U.S. Green Building Council, 2008.

LEED for Neighborhood Development Pilot Rating System: Neighborhood Pattern and Design. Washington, DC: U.S. Green Building Council, 2008.

Summary of Changes to LEED for Homes for Mid-Rise Buildings. Washington, DC: U.S. Green Building Council, 2008.

Practice Exam Part One

Exam time limit: 2 hours

LEED Homes Practice Exam

1. A B C D
2. A B C D E
3. A B C D
4. A B C D E
5. A B C D E
6. A B C D E
7. A B C D E
8. A B C D E
9. A B C D E
10. A B C D E
11. A B C D
12. A B C D
13. A B C D
14. A B C D E
15. A B C D E
16. A B C D E
17. A B C D
18. A B C D
19. A B C D
20. A B C D E
21. A B C D E
22. A B C D E
23. A B C D E
24. A B C D
25. A B C D E
26. A B C D E
27. A B C D
28. A B C D E
29. A B C D
30. A B C D
31. A B C D E
32. A B C D
33. A B C D
34. A B C D E

35. A B C D
36. A B C D E F
37. A B C D
38. A B C D E
39. A B C D
40. A B C D E
41. A B C D
42. A B C D E F
43. A B C D E
44. A B C D E
45. A B C D
46. A B C D
47. A B C D
48. A B C D E
49. A B C D
50. A B C D
51. A B C D E
52. A B C D E F
53. A B C D
54. A B C D E
55. A B C D E
56. A B C D E
57. A B C D E F G
58. A B C D E
59. A B C D
60. A B C D E
61. A B C D E
62. A B C D
63. A B C D E
64. A B C D
65. A B C D
66. A B C D E F
67. A B C D E

68. A B C D
69. A B C D
70. A B C D
71. A B C D E
72. A B C D
73. A B C D
74. A B C D
75. A B C D
76. A B C D E
77. A B C D E
78. A B C D
79. A B C D
80. A B C D
81. A B C D E
82. A B C D
83. A B C D
84. A B C D E F
85. A B C D
86. A B C D E F
87. A B C D
88. A B C D E
89. A B C D E
90. A B C D E
91. A B C D
92. A B C D E
93. A B C D E
94. A B C D
95. A B C D
96. A B C D E F G
97. A B C D E
98. A B C D
99. A B C D E
100. A B C D

Practice Exam Part One

1. Who can view CIRs posted to the USGBC website? (Choose two.)

    (A) individuals with a USGBC website account
    (B) registered employees of USGBC member companies
    (C) LEED Accredited Professionals
    (D) registered project team members

2. LEED project teams can earn an "extra" point for a credit by achieving _____ performance. (Choose two.)

    (A) exemplary
    (B) ideal
    (C) innovative
    (D) original
    (E) perfect

3. Potable water is also called _____.

    (A) drinking water
    (B) graywater
    (C) blackwater
    (D) rainwater

4. When will team member involvement MOST significantly contribute to LEED certification? (Choose three.)

    (A) at conception
    (B) during design development
    (C) during inspector site visits
    (D) throughout the commissioning process
    (E) during construction

5. LEED generally groups credits by credit categories, but the LEED O&M reference guide introduction describes an alternative way of grouping the credits: by their functional characteristics. Which of the following are identified as functional characteristic groups? (Choose two.)

    (A) Administration
    (B) Materials In
    (C) Sustainable Sites
    (D) Waste Management
    (E) Water Efficiency

6. Which of the following actions by a project team will make a LEED project more likely to stay within budget? (Choose three.)

    (A) submitting documentation for as few LEED credits as possible
    (B) adhering to the plan throughout project
    (C) aligning goals with budget
    (D) contacting USGBC for budget guidance
    (E) establishing project goals and expectations

7. Projects in urban areas can utilize which of the following to meet LEED open space requirements? (Choose two.)

    (A) accessible roof decks
    (B) landscaping with indigenous plants
    (C) non-vehicular, pedestrian-oriented hardscapes
    (D) on-site photovoltaics
    (E) pervious parking lots

8. Which of the following can affect a building's energy efficiency? (Choose three.)

    (A) building orientation
    (B) envelope thermal efficiency
    (C) HVAC system size
    (D) refrigerant selection
    (E) VOC content of building materials

9. Which of the following are included in life-cycle cost calculations? (Choose two.)

    (A) equipment
    (B) facility alterations
    (C) maintenance
    (D) occupant transportation
    (E) utilities

10. Which of the following are NOT common benefits of daylighting? (Choose two.)

    (A) increased productivity
    (B) reduced air pollution
    (C) reduced heat island effect
    (D) reduced light pollution
    (E) reduced operating costs

Practice Exam Part One

11. The LEED CI rating system can be applied to which of the following projects? (Choose two.)

    (A) major envelope renovation of a building
    (B) renovation of part of an owner-occupied building
    (C) tenant infill of an existing building
    (D) upgrades to the operation and maintenance of an existing facility

12. Certain prerequisites and credits require project teams to create policies. What information should be included in a policy model? (Choose two.)

    (A) performance period
    (B) policy author
    (C) responsible party
    (D) time period

13. The primary function of which of the following is to encourage sustainable building design, construction, and operation?

    (A) chain-of-custody documentation
    (B) LEED rating systems
    (C) standard operating procedures
    (D) waste reduction program

14. What is required to register a project for LEED certification? (Choose three.)

    (A) list of LEED project team members
    (B) name of LEED AP who will be working on project
    (C) primary contact information
    (D) project owner information
    (E) project type

15. Which of the following are among the five basic steps for pursing LEED for Homes certification? (Choose two.)

    (A) achieve certification as a LEED home
    (B) become a USGBC member company
    (C) create a list of LEED certified products to be used
    (D) include appliances only if they are Energy Star-rated
    (E) market and sell the home

LEED Homes Practice Exam

16. Fire suppression systems that use which of the following will contribute the LEAST to ozone depletion? (Choose two.)

    (A) chlorofluorocarbons
    (B) halons
    (C) hydrochlorofluorocarbons
    (D) hydrofluorocarbons
    (E) water

17. Which of the following describes the *property area*?

    (A) area of the project site impacted by the building and hardscapes
    (B) area of the project site impacted by hardscapes only
    (C) area of the project site impacted by the building only
    (D) total area of the site including constructed and non-constructed areas

18. Which LEED rating system includes performance periods and requires recertification to maintain the building's LEED certification?

    (A) LEED CI
    (B) LEED CS
    (C) LEED EBO&M
    (D) LEED NC

19. For a credit uploaded to LEED Online, a white check mark next to a credit name indicates that the credit is _____.

    (A) being pursued and no online documentation has been uploaded
    (B) being pursued and some online documentation has been uploaded
    (C) complete and ready for submission
    (D) not being pursued

20. Remodeling an older existing building may help a project team achieve credit for building reuse and prevent a project team from achieving credit for _____. (Choose three.)

    (A) controllability of systems
    (B) energy use reduction
    (C) regional materials
    (D) sustainable site use
    (E) water use reduction

21. LEED CS (rather than LEED NC) should be pursued when which of the following items are outside the control of the building owner? (Choose three.)

    (A) envelope insulation
    (B) interior finishes
    (C) lighting
    (D) mechanical distribution
    (E) site selection

22. LEED Online provides a means for _____. (Choose two.)

    (A) code officials to access project documentation
    (B) product vendors to advertize
    (C) project team members to analyze anticipated building energy performance
    (D) project administrators to manage LEED projects
    (E) project team members to manage LEED prerequisites and credits

23. Installing a green roof can help a project team achieve which of the following LEED credits? (Choose two.)

    (A) Development Density and Community Connectivity
    (B) Heat Island Effect
    (C) Light Pollution Reduction
    (D) Site Selection
    (E) Stormwater Design

24. Using lightbulbs with low mercury content, long life, and high lumen output will result in which of the following?

    (A) improved indoor air quality
    (B) increased light pollution
    (C) reduced light pollution
    (D) reduced toxic waste

25. Which of the following are covered in the Sustainable Sites credit category? (Choose three.)

    (A) light to night sky
    (B) light trespass
    (C) on-site renewable energy
    (D) stormwater mitigation
    (E) refrigerants

26. Which of the following tasks must be part of a durability plan? (Choose two.)
    (A) assign responsibilities for plan implementation
    (B) evaluate durability risks of project
    (C) incorporate durability strategies into design
    (D) research local codes
    (E) submit CIR for available list of durability strategies

27. What is the role of the TSAC? (Choose two.)
    (A) ensure the technical soundness of the LEED reference guides and training
    (B) maintain technical rigor and consistency in the development of LEED credits
    (C) resolve issues to maintain consistency across different LEED rating systems
    (D) respond to CIRs submitted by LEED project teams

28. Project teams can earn credit for purchasing sustainable ongoing consumables within the LEED EBO&M rating system. Ongoing consumables containing a defined amount of _____ are considered sustainable. (Choose three.)
    (A) material certified by the Rainforest Alliance
    (B) preindustrial material
    (C) rapidly renewable material
    (D) regionally extracted material
    (E) salvaged material

29. Within an existing building undergoing major renovations, a tenant is pursuing LEED certification. Which rating system should the tenant use to earn a LEED plaque?
    (A) LEED CI
    (B) LEED CS
    (C) LEED EBO&M
    (D) LEED NC

30. Regional Priority credits vary depending on a project's _____.
    (A) certification level
    (B) community connectivity
    (C) development density
    (D) geographic location

31. Which of the following costs should be considered prior to pursuing a LEED credit? (Choose three.)
    (A) application review cost
    (B) construction cost
    (C) documentation cost
    (D) registration cost
    (E) soft cost

32. Which rating system includes CIRs, appeals, and performance periods?

- (A) LEED CI
- (B) LEED CS
- (C) LEED EBO&M
- (D) LEED NC

33. Which standard addresses energy-efficient building design?

- (A) ASHRAE 55
- (B) ANSI/ASHRAE 52.2
- (C) ANSI/ASHRAE 62.1
- (D) ANSI/ASHRAE/IESNA 90.1

34. Which of the following credits will NOT be directly affected by a project team intending to increase a building's ventilation? (Choose two.)

- (A) Controllability of Systems
- (B) Enhanced Commissioning
- (C) Indoor Chemical and Pollutant Source Control
- (D) Measurement and Verification
- (E) Optimize Energy Performance

35. Which of the following is the purpose of a chain-of-custody document?

- (A) to track the movement of products from extraction to the production site
- (B) to track the movement of wood products from the forest to the building site
- (C) to verify rapidly renewable materials
- (D) to verify recycled content

36. A material must be which of the following for it to qualify as a *regional material* within the LEED rating systems? (Choose two.)

- (A) FSC-certified
- (B) made from 10% post-consumer content
- (C) made from products that take 10 years or less to grow
- (D) manufactured within 500 miles of the project site
- (E) permanently installed on the project site
- (F) used to create process equipment

37. A project team's choice of paint will NOT affect which of the following credits?

- (A) Heat Island Effect
- (B) Low-Emitting Materials
- (C) Materials Reuse
- (D) Rapidly Renewable Materials

38. Which of the following do all LEED rating systems contain? (Choose three.)

    (A) core credits
    (B) educational credits
    (C) innovation credits
    (D) operational credits
    (E) prerequisites

39. Which of the following is NOT an example of a durable good?

    (A) computer
    (B) door
    (C) landscaping equipment
    (D) office desk

40. Which of the following would most likely be affected by an increase in ventilation? (Choose two.)

    (A) construction costs
    (B) interior temperature set points
    (C) local climate
    (D) operational costs
    (E) refrigerant management

41. Which items should be considered when selecting refrigerants for a building's HVAC & R system? (Choose two.)

    (A) global depletion potential
    (B) global warming potential
    (C) ozone depletion potential
    (D) ozone warming potential

42. Which information is required to set up a personal account on the USGBC website? (Choose three.)

    (A) company name
    (B) email address
    (C) industry sector
    (D) LEED credentialing exam date
    (E) phone number
    (F) prior LEED projects worked on

43. A project team must comply with which of the following CIRs? (Choose two.)

    (A) CIRs appealed prior to project registration, for all rating systems
    (B) CIRs reviewed by a TAG for the team's own project
    (C) CIRs reviewed by a TAG prior to project registration, for projects within the project's climate region
    (D) CIRs posted prior to project application, for the applicable rating system only
    (E) CIRs posted prior to project completion, for projects within their own project's climate region

44. Which of the following are ways to earn an exemplary performance credit? (Choose two.)

    (A) achieve 75% of the credits in each LEED category
    (B) achieve every LEED credit in the Energy and Atmosphere category
    (C) achieve the next incremental level of an existing credit
    (D) double the requirements of an existing credit
    (E) pursue LEED CI certification within a LEED CS-certified building

45. Which of the following potable water conserving strategies may help a project team achieve a LEED point? (Choose two.)

    (A) collecting blackwater for landscape irrigation
    (B) collecting rainwater for sewage conveyance
    (C) installing an on-site septic tank
    (D) using cooling condensate for cooling tower make-up

46. Which of the following requires an explanation of the proposed credit requirements?

    (A) CIR submittal
    (B) Innovation in Design exemplary performance submittal
    (C) LEED credit equivalence submittal
    (D) LEED Online letter template submittal

47. Which of the following organizations defines the off-site renewable energy sources eligible for LEED credits?

    (A) Center for Research and Development of Green Power
    (B) Center for Resource Solutions
    (C) Department of Energy
    (D) Energy Star

48. The International Code Council (ICC) issues which of the following codes? (Choose three.)

    (A) International Building Automation Code (IBAC)
    (B) International Energy Conservation Code (IECC)
    (C) International Lighting Code (ILC)
    (D) International Mechanical Code (IMC)
    (E) International Plumbing Code (IPC)

49. Which of the following is true?

    (A) A LEED project administrator must be a LEED AP.
    (B) Buildings can be LEED accredited.
    (C) Companies can be USGBC members.
    (D) People can be LEED certified.

50. In addition to satisfying all prerequisites, what is the minimum percentage of points that a project team can earn to achieve LEED Platinum certification?

    (A) 70%
    (B) 80%
    (C) 90%
    (D) 100%

51. LEED submittal templates provide which of the following? (Choose two.)

    (A) a list of potential strategies to achieve a credit or prerequisite
    (B) a means to modify project documentation
    (C) a means to submit a project for review
    (D) a means to review and submit CIRs
    (E) the project's final scorecard

52. According to the *Sustainable Building Technical Manual*, which of the following are key steps of an environmentally responsive design process? (Choose three.)

    (A) bid
    (B) design
    (C) post-design
    (D) pre-design
    (E) rebid
    (F) vendor selection

53. Which is true of the LEED CI rating system?

    (A) Precertification allows the owner to market to potential tenants.

    (B) A project can earn a point for prohibiting smoking within the tenant space.

    (C) A project can earn half points under the Site Selection credit.

    (D) A project must recertify every five years to maintain certification status.

54. Once a project undergoes a construction review, what are the potential rulings for each submitted prerequisite and credit? (Choose three.)

    (A) anticipated

    (B) clarify

    (C) deferred

    (D) denied

    (E) earned

55. What is the role of the LEED Steering Committee? (Choose two.)

    (A) delegate responsibility and oversee all LEED committee activities

    (B) develop LEED accreditation exams

    (C) ensure that LEED and its supporting documentation is technically sound

    (D) establish and enforce LEED direction and policy

    (E) respond to CIRs submitted by LEED project teams

56. The LEED Online workspace allows the LEED project administrator to do which of the following? (Choose three.)

    (A) apply for LEED EBO&M precertification

    (B) assign credits to project team members

    (C) build a project team

    (D) review credit appeals submitted by other project teams

    (E) submit projects for review

57. Which of the following items are free? (Choose three.)

    (A) a project's first CIR

    (B) LEED BD&C reference guide

    (C) LEED brochure

    (D) LEED certification

    (E) LEED for Homes rating system

    (F) sample LEED submittal templates

    (G) usgbc.org account

58. What happens if the scope of a CIR extends beyond the expertise of the assigned TAG? (Choose two.)

    (A) additional response time may be needed
    (B) it is rejected
    (C) it is sent to the LEED Steering Committee
    (D) it must be submitted as an Innovation in Design credit
    (E) the TAG will provide the ruling

59. What becomes available or accessible once a project is registered?

    (A) LEED project tools
    (B) posted CIRs
    (C) posted credit appeals
    (D) sample submittal templates

60. The LEED rating systems require compliance with which of the following? (Choose two.)

    (A) codes and regulations that address asbestos and water discharge
    (B) codes and regulations that address PCBs and water management
    (C) referenced standards that address fixture performance requirements for water use
    (D) referenced standards that address sustainable forest management practices
    (E) referenced standards that address VOCs

61. Which of the following steps are part of creating an integrated project team? (Choose two.)

    (A) include members from varying industry sectors
    (B) include product vendors in the design phase
    (C) involve a commissioning authority in team members' selection
    (D) involve the LEED AP in design integration
    (E) involve the team in different project phases

62. Facilities undergoing minor alterations and system upgrades must follow which rating system?

    (A) LEED CI
    (B) LEED CS
    (C) LEED EBO&M
    (D) LEED NC

63. Which of the following strategies may help a project team achieve a LEED credit? (Choose two.)

    (A) establish an erosion and sedimentation plan
    (B) establish a location for the storage and collection of recyclables
    (C) install HCFC-based HVAC & R equipment
    (D) use rainwater for sewage conveyance or landscape irrigation
    (E) remediate contaminated soil

64. Which of the following statements is true?

    (A) CIRs are reviewed by a TAG at no additional charge once a project is registered.
    (B) The first step toward LEED certification is passing the Green Associate exam.
    (C) The LEED rating system only applies to commercial buildings.
    (D) To achieve LEED certification, every prerequisite and a minimum number of points must be earned.

65. Installing ground source heat pumps would help a project team achieve which of the following credits or prerequisites?

    (A) Green Power
    (B) Optimize Energy Performance
    (C) On-Site Renewable Energy
    (D) Water Use Reduction

66. What are the goals of the Portfolio Program? (Choose two.)

    (A) to create a volume accreditation path
    (B) to encourage global adoption of sustainable green building practices
    (C) to exceed the requirements of ANSI/ASHRAE/IESNA 90.1 by a rating system-designated percentage
    (D) to offer a volume certification path
    (E) to provide a streamlined certification process for large-scale projects
    (F) to provide a template of key data for the design team members to compile

67. When should a LEED project's budget be addressed? (Choose two.)

    (A) before the design phase
    (B) during the construction phase
    (C) during the selection of construction team
    (D) during the selection of design team
    (E) upon project completion

68. Which of the following could help minimize potable water used for the site landscaping and contribute toward earning a LEED landscaping credit?

    (A) designing the site or the building's roof with no landscaping
    (B) installing invasive plants
    (C) installing native or adapted plants
    (D) installing turf grass

69. How many Regional Priority points are available for LEED projects?

    (A) 2 points
    (B) 4 points
    (C) 6 points
    (D) 8 points

70. A project team chooses to group credits by functional characteristics. Credits focusing on the measurement of a building's energy performance and ozone protection would be grouped into which of the following categories?

    (A) Energy and Atmosphere
    (B) Energy Metrics
    (C) Materials Out
    (D) Site Management

71. Which of the following are considered principle durability risks? (Choose three.)

    (A) heat islands
    (B) interior moisture loads
    (C) ozone depletion
    (D) pests
    (E) ultraviolet radiation

72. What is the first step of the LEED for Homes certification process?

    (A) become a LEED AP
    (B) contact a LEED for Homes provider
    (C) register with GBCI
    (D) submit a CIR

73. HVAC & R equipment with _____ will contribute to ozone depletion and global warming.

    (A) a manufacture date prior to 2005
    (B) a relatively long equipment life
    (C) minimal refrigerant charge
    (D) refrigerant leakage

74. Sealing ventilation ducts, installing rodent- and corrosion-proof screens, and using air-sealing pump covers are strategies that could be a part of which of the following?

    (A) design charrette
    (B) durability plan
    (C) landscape management plan
    (D) PE exemption form

75. Which standard addresses the thermal comfort of building occupants?

    (A) ASHRAE 55
    (B) ANSI/ASHRAE 52.2
    (C) ANSI/ASHRAE 62.1
    (D) ANSI/ASHRAE/IESNA 90.1

76. A LEED project's primary contact must submit which of the following when registering a project? (Choose three.)

    (A) email address
    (B) LEED AP certificate
    (C) LEED project history
    (D) organization name
    (E) individual's title

77. Prior to registering a LEED project, which of the following must be confirmed? (Choose two.)

    (A) necessary CIRs
    (B) precertification
    (C) project cost
    (D) project summary
    (E) project team members

78. Water lost through plant transpiration and evaporation from soil is described by the term _____.

    (A) evapotranspiration
    (B) infiltration
    (C) sublimation
    (D) surface runoff

79. Individuals with a LEED reference guide electronic access code can do which of the following?

    (A) join a LEED project as a team member
    (B) print the LEED reference guide
    (C) purchase a LEED reference guide
    (D) view a protected electronic version of the LEED reference guide

80. LEED credits and prerequisites are presented in a common format in all versions of LEED rating systems. This format includes which of the following?

    (A) economic impact
    (B) greening opportunities
    (C) intent
    (D) submittal requirements

81. The Materials and Resources category directly addresses which of the following? (Choose two.)

    (A) habitat conservation
    (B) durable goods
    (C) energy consumption
    (D) landscape
    (E) waste stream

82. Building on which of the following sites will most likely have the smallest impact on the environment?

    (A) greenfield
    (B) public parkland
    (C) previously undeveloped site
    (D) urban area

83. Every LEED submittal template requires which of the following items?

    (A) declarant's name
    (B) product manufacturer
    (C) project area
    (D) project location

84. The pre-design phase of a LEED project should include which of the following steps? (Choose three.)

    (A) commissioning mechanical systems
    (B) establishing a project budget
    (C) establishing project goals
    (D) site selection
    (E) testing and balancing mechanical systems
    (F) training maintenance staff

85. Which is true about a CIR submittal?

    (A) A CIR must be submitted as a text-based inquiry.
    (B) Drawings and specification sheets must be submitted as attachments.
    (C) It must include a complete project narrative.
    (D) Text is limited to 1000 words.

86. Green building design and construction decisions should be guided by which of the following items? (Choose three.)

   (A) bid cost
   (B) construction documents
   (C) design cost
   (D) energy efficiency
   (E) environmental impact
   (F) indoor environment

87. GBCI is a nonprofit organization that provides which of the following services? (Choose two.)

   (A) accreditation of industry professionals
   (B) certification of sustainable products
   (C) certification of sustainable buildings
   (D) educational programs on sustainability topics

88. Credits can be earned after submitting which of the following? (Choose two.)

   (A) appeal documentation
   (B) design documentation
   (C) certification documentation
   (D) construction documentation
   (E) post-construction documentation

89. What are the short-term benefits of commissioning? (Choose two.)

   (A) assured credit achievement
   (B) decreased initial project cost
   (C) promotion of code compliance
   (D) promotion of design efficiency
   (E) reduced design and construction time

90. Conventional fossil-based electricity generation results in which of the following emissions? (Choose three.)

   (A) anthropogenic nitrogen oxide
   (B) carbon dioxide
   (C) carbon monoxide
   (D) sulfur dioxide
   (E) VOCs

91. Which strategy helps minimize a site's heat island effect?

    (A) having a high glazing factor
    (B) installing hardscapes with low SRI values
    (C) maximizing the area of site hardscapes
    (D) shading hardscapes with vegetation

92. The BOD, which includes design information necessary to accomplish the owner's project requirements, must contain which of the following? (Choose three.)

    (A) building materials selection
    (B) indoor environmental quality criteria
    (C) mechanical systems descriptions
    (D) process equipment energy consumption information
    (E) references to applicable codes

93. An integrated project team should include which of the following professionals? (Choose two.)

    (A) code official
    (B) energy sustainability consultant
    (C) landscape architect
    (D) product manufacturer
    (E) utility manager

94. A project is considered a major renovation when at least _____ of the building envelope, interior, or mechanical systems is modified.

    (A) 50%
    (B) 60%
    (C) 70%
    (D) 80%

95. A building that will be partially occupied by the owner may pursue LEED CS certification if the building owner will occupy no more than _____ of the building's leasable space.

    (A) 25%
    (B) 50%
    (C) 75%
    (D) 80%

96. The LEED EBO&M rating system includes a Best Management Practices prerequisite. This prerequisite would most likely fall under which of the following credit groupings? (Choose two.)

    (A) Energy and Atmosphere
    (B) Indoor Environmental Quality
    (C) Innovation in Design
    (D) Materials and Resources
    (E) Occupant Health and Productivity
    (F) Operational Effectiveness
    (G) Site Management

97. The Project Details section of the LEED project registration form requires which of the following? (Choose three.)

    (A) company names of all team members
    (B) gross area of the building
    (C) list of likely innovation credits to be pursued
    (D) project budget
    (E) site conditions

98. To be eligible for LEED recertification, a project must be

    (A) precertified under the LEED for Homes rating system
    (B) previously certified under the LEED EBO&M rating system
    (C) previously certified as LEED Platinum under the LEED CS rating system
    (D) previously certified at any level other than LEED Platinum under the LEED NC rating system

99. Which of the following can help reduce a building's energy load? (Choose two.)

    (A) reducing the building's heat island effect
    (B) increasing the ventilation rate
    (C) installing heat recovery systems
    (D) flushing out prior to occupancy
    (E) zoning mechanical systems

100. Minimizing which of the following will improve a building's indoor environmental quality?

    (A) acoustical control
    (B) natural lighting
    (C) ventilation rates
    (D) VOC content in building materials

Practice Exam Part Two

Exam time limit: 2 hours

LEED Homes Practice Exam

1. A B C D E
2. A B C D
3. A B C D E
4. A B C D
5. A B C D E
6. A B C D E
7. A B C D E
8. A B C D E
9. A B C D
10. A B C D E
11. A B C D
12. A B C D
13. A B C D
14. A B C D E
15. A B C D E
16. A B C D E
17. A B C D
18. A B C D
19. A B C D
20. A B C D
21. A B C D
22. A B C D E
23. A B C D E F
24. A B C D E F
25. A B C D
26. A B C D E
27. A B C D
28. A B C D
29. A B C D
30. A B C D E
31. A B C D
32. A B C D E
33. A B C D E
34. A B C D
35. A B C D
36. A B C D E
37. A B C D E
38. A B C D
39. A B C D E
40. A B C D
41. A B C D E
42. A B C D E
43. A B C D E
44. A B C D
45. A B C D
46. A B C D
47. A B C D
48. A B C D E
49. A B C D
50. A B C D E
51. A B C D
52. A B C D
53. A B C D
54. A B C D
55. A B C D E
56. A B C D E F
57. A B C D
58. A B C D
59. A B C D
60. A B C D
61. A B C D E F
62. A B C D
63. A B C D
64. A B C D
65. A B C D E
66. A B C D E
67. A B C D
68. A B C D
69. A B C D
70. A B C D
71. A B C D
72. A B C D
73. A B C D
74. A B C D E F
75. A B C D E
76. A B C D
77. A B C D
78. A B C D E F
79. A B C D
80. A B C D
81. A B C D E
82. A B C D E
83. A B C D
84. A B C D E
85. A B C D
86. A B C D
87. A B C D
88. A B C D
89. A B C D
90. A B C D E F
91. A B C D E
92. A B C D
93. A B C D
94. A B C D E
95. A B C D
96. A B C D
97. A B C D E
98. A B C D E F
99. A B C D E
100. A B C D

Practice Exam Part Two

1. Which of the following are NOT required of the one-hour walk-through for AE Prerequisite 1.1, Education of the Homeowner or Tenant: Basic Operations Training? (Choose two.)

    (A) discussion of furniture and appliance selections
    (B) explanation of the LEED Scorecard
    (C) identification of all installed equipment
    (D) instruction on equipment operation
    (E) information on equipment maintenance

2. How many points can a project earn in the Locations and Linkages credit category?

    (A) 0 points
    (B) 10 points
    (C) 13 points
    (D) 20 points

3. Building on a previously undeveloped site with which soils will preclude a home from LEED certification? (Choose three.)

    (A) alfisols
    (B) prime soils
    (C) unique soils
    (D) ustic soils
    (E) soils of state significance

4. What is the minimum reduction in irrigation demand required for a home to earn points for the irrigation-related Water Efficiency and Sustainable Sites credits?

    (A) 30%
    (B) 35%
    (C) 40%
    (D) 45%

5. To meet the requirements of ID Credit 2, Durability Management Process, water-resistant flooring must be installed in which rooms? (Choose two.)

    (A) basement
    (B) garage
    (C) kitchen
    (D) laundry room
    (E) tank water heater room

6. A green rater is responsible for conducting performance tests for which of the following? (Choose three.)

    (A) duct leakage
    (B) indoor air quality
    (C) HVAC refrigerant charge
    (D) local exhaust
    (E) outdoor air flow

7. As described in AE Prerequisite 1.1, Education of the Homeowner or Tenant: Basic Operations Training, which of the following should NOT be included in a LEED home's operations and maintenance manual? (Choose two.)

    (A) durability inspection checklist
    (B) irrigation system inspection checklist
    (C) product information for installed equipment
    (D) signed accountability form
    (E) thermal bypass checklist

8. Installing low-flow fixtures in all bathrooms would help a home meet the requirements of which credits? (Choose two.)

    (A) WE Credit 3, Indoor Water Use
    (B) EA Credit 1, Optimize Energy Performance
    (C) EA Credit 7, Water Heating
    (D) SS Credit 4, Surface Water Management
    (E) EQ Credit 3, Moisture Control

9. A LEED home with photovoltaic panels supplying 2000 kWh of renewable energy annually has an actual annual reference electricity load of 8000 kWh. The corresponding HERS reference home has an annual reference electricity load of 10,000 kWh. For this strategy, how many points can the home earn for EA Credit 10, Renewable Energy?

    (A) 0 points
    (B) 2 points
    (C) 6 points
    (D) 10 points

10. Which strategies contribute to meeting the requirements of EQ Credit 8.2, Contaminant Control: Indoor Contaminant Control? (Choose two.)

    (A) flushing the home for 48 hours
    (B) having a detached garage
    (C) installing a central vacuum system
    (D) installing permanent walk-off mats
    (E) sealing all permanent ducts and vents during construction

11. Which credit specifies emissions standards for VOCs?

    (A) EQ Credit 8, Contaminant Control

    (B) MR Credit 2, Environmentally Preferable Products

    (C) EA Prerequisite 11, Residential Refrigerant Management

    (D) AE Credit 2, Education of Building Manager

12. What organization is responsible for maintaining the International Energy Conservation Code's Climate Zone Map?

    (A) AIA

    (B) NOAA

    (C) U.S. EPA

    (D) U.S. DOE

13. What is the building code that one- and two-family dwellings of three stories or less must follow?

    (A) National Building Code

    (B) International Building Code

    (C) International Residential Code

    (D) Federal Building Code

14. Which of the following is NOT considered a rapidly renewable material? (Choose two.)

    (A) agrifiber

    (B) cork

    (C) linoleum

    (D) pine

    (E) soapstone

15. Which of the following items are listed in the Energy Star Thermal Bypass Inspection Checklist? (Choose three.)

    (A) attic ceiling

    (B) attic eave vents

    (C) common walls between dwelling units

    (D) garage slab

    (E) walls adjoining conditioned spaces

16. Installing which of the following will NOT reduce surface water runoff from the site? (Choose two.)

    (A) bioswale

    (B) graywater system

    (C) high-efficiency irrigation system

    (D) permeable paving

    (E) rain garden

17. A home is located within a half mile of seven basic community resources and transit services that offer 125 transit rides or more per weekday. How many points would the home qualify for under LL Credit 5, Community Resources/Transit?

    (A) 0 points
    (B) 1 point
    (C) 2 points
    (D) 3 points

18. Which of the following is NOT a feature of a high-efficiency irrigation system?

    (A) central shut-off valve
    (B) low-flow fittings
    (C) moisture sensor controller
    (D) pressure-regulating device

19. *U-value* is a measure of which of the following?

    (A) airflow
    (B) humidity
    (C) thermal conductivity
    (D) thermal resistance

20. Which of the following organizations is responsible for regulating radon exposure?

    (A) AIA
    (B) ASHRAE
    (C) ASTM
    (D) U.S. EPA

21. To qualify for LEED certification, what is the minimum MERV rating for a home's air filters?

    (A) MERV 8
    (B) MERV 11
    (C) MERV 13
    (D) MERV 15

22. Installing which of the following reduces the energy needed for heating water? (Choose three.)

    (A) efficient appliances
    (B) a high-efficiency irrigation system
    (C) low-flow showerheads
    (D) pipe insulation
    (E) a rainwater harvesting system

23. Graywater is NOT collected from which of the following sources? (Choose three.)

    (A) clothes washers
    (B) dishwashers
    (C) showers
    (D) toilets
    (E) municipal wells
    (F) bathroom sinks

24. Which of the following are NOT referenced in EA Credit 2, Insulation? (Choose two.)

    (A) ACCA's *Manual J*
    (B) ASHRAE's *Handbook of Fundamentals*
    (C) U.S. EPA's Energy Star Structural Insulated Panel Visual Inspection Form
    (D) U.S. EPA's Energy Star Thermal Bypass Inspection Checklist
    (E) ICC's International Energy Conservation Code
    (F) RESNET's HERS Index

25. Which of the following is NOT necessary for compliance with SS Credit 2.2, Basic Landscape Design?

    (A) Compacted soil must be tilled to at least 6 inches.
    (B) The landscape design must include drought-tolerant plants.
    (C) Turf must not be used in densely-shaded areas.
    (D) Installed turf must be drought-tolerant.

26. Which of the following help determine the size of HVAC equipment needed in a home? (Choose three.)

    (A) the air tightness of the home's envelope
    (B) the home's receptacle load
    (C) the quality of the home's insulation
    (D) the home's size
    (E) the type of flooring material installed in the home

27. In which state will the solar heat gain coefficient requirement be MOST stringent for a home with a total window-to-floor area ratio of 20%?

    (A) Iowa
    (B) Pennsylvania
    (C) Florida
    (D) Utah

28. Which organization or agency created the indoor air quality package referenced in LEED for Homes?

    (A) ASHRAE
    (B) U.S. EPA
    (C) Greenguard Environmental Institute
    (D) Green Seal

29. Which of the following is NOT considered composite wood?

    (A) cork
    (B) oriented-strand board
    (C) plywood
    (D) wheatboard

30. Which of the following must be Energy Star-rated to meet the requirements of EA Credit 8.3, Advanced Lighting Package? (Choose two.)

    (A) 25% of emergency lighting
    (B) 50% of hallway lights
    (C) 60% of hard-wired fixtures
    (D) 100% of ceiling fans
    (E) 100% of closet lights

31. For which of the following systems may a home require a special permit?

    (A) drip irrigation system
    (B) efficient HVAC system
    (C) graywater reuse system
    (D) security system

32. Green Seal Standard GS-11 sets VOC limits for which of the following? (Choose two.)

    (A) adhesives
    (B) carpets
    (C) paints
    (D) primers
    (E) stains

33. Installing Energy Star-rated ceiling fans could help a project meet the requirements of which of the following prerequisites or credits? (Choose two.)

    (A) EQ Credit 1, Energy Star with Indoor Air Package
    (B) EA Prerequisite 3, Air Infiltration
    (C) EA Prerequisite 6, Space Heating and Cooling Equipment
    (D) EA Credit 8, Lighting
    (E) EA Credit 9, Appliances

34. Which professional is approved to use HERS software?

    (A) electrical engineers only
    (B) HERS-trained energy raters only
    (C) green raters and HERS-trained energy raters
    (D) LEED APs and electrical engineers

35. Which of the following wetlands would be EXEMPT from the site selection requirements of LL Credit 2, Site Selection?

    (A) a natural wetland less than 10 years old
    (B) a wetland within 100 feet of the home site
    (C) an artificial wetland used for stormwater mitigation
    (D) a wetland that is above the floodplain

36. Which of the following are conveyance system components for a rainwater harvesting system? (Choose two.)

    (A) cistern
    (B) downspout
    (C) gutter
    (D) pump
    (E) tank

37. Which of the following are NOT considered buildable land according to LL Credit 2, Site Selection? (Choose two.)

    (A) land 15 feet above FEMA's 100-year floodplain
    (B) open space protected by local code
    (C) previously developed land
    (D) public parkland
    (E) undeveloped land with native vegetation

38. What does permeability measure?

    (A) a material's ability to allow liquid to pass through it
    (B) a liquid's ability to penetrate a material
    (C) a liquid's rate of evaporation from a material
    (D) a liquid's rate of runoff from a material

39. Which of the following are NOT required of EQ Prerequisite 5.1, Basic Local Exhaust? (Choose two.)

    (A) Continuously operate the exhaust fan.
    (B) Exhaust air to the outdoors.
    (C) Install an exhaust fan in the attic.
    (D) Install Energy Star-rated exhaust fans.
    (E) Meet air flow requirements of ANSI/ASHRAE 62.2.

40. A home in the International Energy Conservation Code's climate zone 1 with a HERS rating of 95 will be eligible to earn how many points for EA Credit 1?

    (A) 0 points
    (B) 1 point
    (C) 6 points
    (D) 10 points

41. Which strategies contribute to meeting the requirements of MR Credit 1, Material-Efficient Framing? (Choose two.)

    (A) preparing detailed framing plans
    (B) recycling all excess framing material
    (C) using FSC-certified wood
    (D) using off-site fabrication
    (E) using salvaged wood

42. A home must have which of the following to meet basic combustion venting requirements of EQ Prerequisite 2.1, Basic Combustion Venting? (Choose two.)

    (A) carbon monoxide monitors on each floor
    (B) doors on all fireplaces
    (C) no fireplaces
    (D) unvented combustion appliances
    (E) wood burning stoves for heating

43. Which system requires R-3 insulation or better around distribution pipes in unconditioned areas?

    (A) central manifold system
    (B) forced air system
    (C) hydronic system
    (D) plumbing system

44. Which of the following are benefits of compact land development? (Choose three.)

    (A) increased mass transit use
    (B) increased pedestrian activity
    (C) protection of endangered species
    (D) protection of undeveloped land
    (E) reduced irrigation demand

45. How many gallons of water per year must a graywater system reuse in order to earn points for WE Credit 1.2, Water Reuse: Graywater Reuse System?

    (A) 2000 gallons
    (B) 5000 gallons
    (C) 7500 gallons
    (D) 10,000 gallons

46. What organization manages the Green Label Plus program?

    (A) U.S. EPA
    (B) ASTM
    (C) Carpet and Rug Institute
    (D) Green Seal

47. For a home with a forced-air system to meet the requirements of EA Credit 5.3, Heating and Cooling Distribution: Minimal Distribution Losses, what is the maximum duct leakage allowed per minute at 25 Pa per 100 sq ft?

    (A) 1.0 cu ft
    (B) 2.0 cu ft
    (C) 3.0 cu ft
    (D) 4.0 cu ft

48. Which of the following are NOT appropriate storage mediums for harvested rainwater? (Choose three.)

    (A) cistern
    (B) constructed wetland
    (C) pond
    (D) rain garden
    (E) tank

49. What does U.S. EPA's WaterSense program certify individuals to install?

    (A) boilers
    (B) irrigation systems
    (C) graywater systems
    (D) fixtures and fittings

50. Which of the following statements are true of fly ash? (Choose two.)

    (A) It is a byproduct of burning coal.
    (B) It is sometimes a component of concrete.
    (C) It is sometimes a component of particle board.
    (D) It is a type of mulch.
    (E) It is not recyclable.

51. What is ACCA's *Manual D* used for?

    (A) selecting a SEER number
    (B) sizing HVAC systems
    (C) sizing duct systems
    (D) sizing return grills

52. What does SEER measure the energy efficiency of?

    (A) a home's air conditioning system
    (B) a home's air filtration system
    (C) a home's lighting system
    (D) an entire home

53. What organization offers chain-of-custody certification?

    (A) Forest Stewardship Council
    (B) National Resources Conservation Service
    (C) National Fenestration Rating Council
    (D) Residential Energy Services Network

54. What is albedo a measure of?

    (A) solar radiation
    (B) surface reflectivity
    (C) solar heat gain
    (D) thermal resistance

55. Which of the following are considered post-consumer waste? (Choose three.)

    (A) demolition debris
    (B) grass clippings
    (C) material collected through recycling programs
    (D) industrial metal trimmings
    (E) timber processing sawdust

56. Which of the following plant types are BEST suited for xeriscaping? (Choose three.)

    (A) drought-tolerant plants
    (B) aquatic plants
    (C) native plants
    (D) noxious plants
    (E) herbaceous plants
    (F) indigenous plants

57. How many climate zones has the International Energy Conservation Code defined in the U.S.?

    (A) 4 climate zones
    (B) 6 climate zones
    (C) 8 climate zones
    (D) 10 climate zones

58. What can ACCA's *Manual J* help a project team with?

    (A) designing a duct system
    (B) calculating heating and cooling loads
    (C) evaluating the design and construction of a masonry heater
    (D) measuring the energy performance of windows

59. How many points can a project earn for installing two Energy Star refrigerators, an Energy Star dishwasher, and an Energy Star clothes washer in a home?

    (A) 1 points
    (B) 2 points
    (C) 3 points
    (D) 4 points

60. Which of the following is a rain garden MOST like?

    (A) cistern
    (B) constructed wetland
    (C) pond
    (D) vegetated swale

61. LEED for Homes encourages the use of which types of water for irrigation? (Choose three.)

    (A) blackwater
    (B) graywater
    (C) harvested rainwater
    (D) municipal water
    (E) potable water
    (F) reclaimed water

62. What is the maximum relative humidity percentage permitted for a home to meet the requirements of EQ Credit 3, Moisture Control?

    (A) 40%
    (B) 50%
    (C) 60%
    (D) 70%

63. According to the home size adjustment calculation in the LEED for Homes rating system, what is the maximum area for a two bedroom home to be considered "neutral"?

    (A) 900 sq ft
    (B) 1400 sq ft
    (C) 1900 sq ft
    (D) 2200 sq ft

64. To comply with Energy Star requirements used in LEED for Homes, third-party testing is NOT required for which of the following?

    (A) back-draft potential
    (B) blower doors
    (C) envelope leakage performance
    (D) thermal bypass

65. Which of the following generate renewable energy, as defined by USGBC? (Choose three.)

    (A) biofuel systems
    (B) geothermal systems
    (C) passive solar heat systems
    (D) photovoltaic systems
    (E) wood furnaces

66. Under what circumstances can private land be considered public open space? (Choose two.)

    (A) if public events are hosted annually on the land
    (B) if the land has deeded public access
    (C) if there is a bike trail through the land
    (D) if there is a precedent of public use and a 10 year commitment to future public use
    (E) if there is pedestrian access through the land

67. Who sets the guidelines for the Energy Star program?

    (A) USDA
    (B) U.S. DOE
    (C) U.S. EPA
    (D) USGBC

68. Which of the following is NOT a requirement of EQ Credit 8.2, Contaminant Control: Indoor Contaminant Control?

    (A) cleaning the HVAC filter after the preoccupancy flush
    (B) flushing the home for 48 hours prior to occupancy
    (C) keeping all interior doors open during the preoccupancy flush
    (D) turning on all dehumidification equipment during the preoccupancy flush

69. For a home with 2000 sq ft of conditioned floor area, what is the maximum area of skylight glazing area permissible to comply with the requirements EA Credit 4.1, Good Windows?

    (A) 20 sq ft
    (B) 30 sq ft
    (C) 40 sq ft
    (D) 60 sq ft

70. Where does a product have to be extracted, processed, and manufactured in order to qualify as local or regional within the Materials and Resources credit category?

    (A) within 500 miles of the home
    (B) within the same climate zone as the home
    (C) within the same geographical region as the home
    (D) within the same state as the home

71. Which of the following is NOT required for LEED for Homes certification?

    (A) durability planning
    (B) fundamental commissioning
    (C) green rater verification
    (D) waste factor reduction

72. The area of which of the following is NOT included in the "built environment" calculations for SS Credit 4.1, Surface Water Management: Permeable Lot?

    (A) driveway
    (B) garden
    (C) house footprint
    (D) walkway

73. LL Credit 3, Preferred Locations, describes what term as a development of new homes adjacent to an established community?

    (A) edge development
    (B) high-density development
    (C) infill development
    (D) low-impact development

74. Wood grown in which of the following locations can qualify for MR Prerequisite 2.1, Environmentally Preferable Products: FSC Certified Tropical Wood? (Choose two.)

    (A) California
    (B) Florida
    (C) Nicaragua
    (D) Philippines
    (E) South Africa
    (F) Spain

75. Which of the following projects is NOT eligible for LEED certification under the LEED for Homes rating system?

    (A) a five-story multifamily condominium
    (B) a luxury duplex
    (C) a modular home
    (D) an affordable three-story multifamily building

76. For SS Credit 6, Compact Development, the building density of a residential area is determined using which of the following?

    (A) the floor area per acre of buildable land
    (B) the number of basic services per half mile
    (C) the number of bus lines per quarter mile
    (D) the number of dwelling units per acre of buildable land

77. Which of the following energy reduction strategies must be implemented the EARLIEST in the design process?

    (A) minimizing air leakage for conditioned spaces
    (B) minimizing east-west sun exposure
    (C) maximizing insulation
    (D) minimizing thermal bridges

78. USGBC defines which of the following terms as building components recovered from a demolition site and reused in their original state? (Choose three.)

    (A) construction waste
    (B) preconsumer material
    (C) reclaimed material
    (D) recycled material
    (E) reused material
    (F) salvaged material

79. A material with a high SRI value will also have a high value for which of the following?

    (A) albedo
    (B) daylight factor
    (C) glazing factor
    (D) visible light transmittance

80. What is NOT true of sedimentation?

    (A) It is caused by stormwater runoff.
    (B) It causes the aging of water bodies.
    (C) It is different from siltation.
    (D) It causes erosion.

81. Installing which of the following will NOT comply with framing efficiency requirements of MR Credit 1.4, Material-Efficient Framing: Framing Efficiencies? (Choose two.)

    (A) three-stud corners
    (B) floor joist spacing at 16 inches on center
    (C) open web floor trusses
    (D) structural insulated panel (SIP) walls
    (E) stud spacing at 24 inches on center

82. AE Credit 2, Education of Building Manager, requires building managers of dwellings with more than five units to provide guidance to occupants for which of the following? (Choose three.)

    (A) appliance selection
    (B) furniture selection
    (C) landscaping
    (D) lighting selection
    (E) unit size selection

83. For compliance with LL Credit 2, Site Selection, when is it permissible to build a LEED for Homes project on a soil of state significance?

    (A) if a qualified member of the project team recommended it
    (B) if the builder agrees to construct a wetland on the site
    (C) if the site has been previously developed
    (D) if there are no endangered species on the site

84. Which of the following strategies are used to manage stormwater? (Choose three.)

    (A) minimizing the area of impermeable hardscape
    (B) increasing the under-roof area
    (C) installing a vegetated roof
    (D) installing drought-tolerant plants
    (E) installing permeable paving

85. According to USGBC, which of the following is NOT made from reclaimed material?

    (A) a deck made from milled wood found in dumpsters
    (B) a table made from cedar felled by a storm
    (C) stair treads made from floor joists
    (D) garage stools refinished and used as barstools

86. What is the maximum number of points that a LEED for Homes project can earn in the Energy and Atmosphere credit category?

    (A) 16 points
    (B) 21 points
    (C) 22 points
    (D) 38 points

87. Which of the following is NOT usually a benefit of a Planned Unit Development?

    (A) increased open space preservation
    (B) increased site design efficiency
    (C) reduced home construction costs
    (D) reduced infrastructure costs

88. What is the minimum efficiency of a high-efficiency toilet?

    (A) 1.0 gallons of water per flush
    (B) 1.3 gallons of water per flush
    (C) 1.5 gallons of water per flush
    (D) 2.0 gallons of water per flush

89. Where is an aerator typically installed?

    (A) in a cistern
    (B) in a faucet
    (C) in a hydronic system
    (D) in HVAC equipment

90. If a project earns points for EA Credit 1, Optimize Energy Performance, for which of the following credits can it also earn points? (Choose two.)

    (A) EA Credit 2.2, Enhanced Insulation
    (B) EA Credit 7.2, Water Heating: Pipe Insulation
    (C) EA Credit 7.3, Water Heating: Efficient Domestic Hot Water Equipment
    (D) EA Credit 9.1, High-Efficiency Appliances
    (E) EA Credit 8.2, Improved Lighting
    (F) EA Credit 11.2, Residential Refrigerant Management: Appropriate HVAC Refrigerants

91. Which of the following are considered hardscape elements? (Choose three.)

    (A) fountains
    (B) gravel
    (C) paving stones
    (D) ponds
    (E) trees

92. Which of the following is used to determine if the insulation level of a home complies with the International Energy Conservation Code?

    (A) blower door test
    (B) HERS index
    (C) pre-drywall inspection
    (D) REScheck

93. Which of the following factors is NOT used to calculate the landscape coefficient?

    (A) density factor
    (B) evapotranspiration factor
    (C) microclimate factor
    (D) species factor

94. Why does USGBC encourage compact land development? (Choose three.)

    (A) to protect endangered species
    (B) to reduce irrigation demand
    (C) to promote mass transit use
    (D) to promote pedestrian activity
    (E) to protect undeveloped land

95. What is the maximum HERS index for a home to earn points for EA Credit 1, Optimize Energy Performance?

    (A) 64
    (B) 79
    (C) 84
    (D) 99

96. EA Credit 11.2, Residential Refrigerant Management: Appropriate HVAC Refrigerants, allows the use of which of the following refrigerants?

    (A) CFC
    (B) HCFC
    (C) R22
    (D) R410a

97. What must a home's forced air heating and cooling distribution system have to comply with the requirements of EA Credit 5, Heating and Cooling Distribution System? (Choose two.)

    (A) R-3 insulation around ducts in unconditioned spaces
    (B) hard ducts on all returns
    (C) insulated distribution pipes
    (D) sealed duct seams
    (E) controlled water distribution temperature

98. Which of the following reduces heat island effect? (Choose three.)

    (A) increasing the shaded area
    (B) installing drought-tolerant plants
    (C) installing white concrete
    (D) installing open pavers
    (E) limiting conventional turf use
    (F) reducing light pollution

99. The central manifold distribution system must meet requirements for which of the following in order to be eligible for 2 points under EA Credit 7.1, Efficient Hot Water Distribution? (Choose two.)

    (A) air leakage
    (B) branch line diameter
    (C) distribution losses
    (D) refrigerant charge
    (E) trunk length

100. Which of the following strategies will NOT earn a home points within the Sustainable Sites category?

    (A) choosing a compact lot
    (B) choosing a previously developed lot
    (C) controlling pests
    (D) reducing irrigation demand

Practice Exam Part One Answers

Answers begin on the page that follows.

LEED Homes Practice Exam — Answer Sheet

1. B, D
2. A, C
3. A
4. A, B, E
5. A, B
6. B, C, E
7. A, C
8. A, B, C
9. C, E
10. C, D
11. B
12. C, D
13. B
14. C, D, E
15. A, E
16. D, E
17. D
18. C
19. B
20. B, D, E
21. B, C, D
22. D, E
23. A
24. D
25. A, B, D
26. B, C
27. A, C
28. B, C
29. A
30. D
31. B, C, E
32. C
33. D
34. A, C
35. B
36. D, E
37. C
38. A, C
39. B
40. C, D
41. C
42. B, C, E
43. B, D
44. C
45. C
46. C
47. B
48. B, D
49. C
50. C
51. C, E
52. A, B
53. C
54. B, D
55. A, D
56. B, C
57. E, F, G
58. A, C
59. A
60. A, B
61. A, E
62. C
63. D, E
64. D
65. B
66. D, E
67. A, B
68. C
69. B
70. B
71. D, E
72. B
73. D
74. B
75. A
76. A, D, E
77. D, E
78. A
79. D
80. C
81. B, E
82. D
83. A
84. B, C, D
85. A
86. D, E, F
87. A, C
88. A, D
89. D, E
90. A, B, E
91. D
92. B, C, E
93. B
94. A
95. B
96. A, F
97. B, D, E
98. B
99. C, E
100. D

Practice Exam Part One Answers

1. *The answers are:* **(B)** registered employees of USGBC member companies
   (D) registered project team members

Only members of a LEED project team and employees of a USGBC member company may view posted CIRs.

2. *The answers are:* **(A)** exemplary
   (C) innovative

To earn points in the Innovation in Design category, a project can exceed the requirements of an existing LEED credit and earn exemplary performance points, or the project team can implement innovative performance strategies not addressed by the LEED rating systems.

3. *The answer is:* **(A)** drinking water

Potable water is water that meets or exceeds EPA drinking water standards and is supplied from wells or municipal water systems.

4. *The answers are:* **(A)** at conception
   (B) during design development
   (E) during construction

LEED projects benefit from the inclusion of team members in ongoing commissioning and inspector site visits; however, team member participation is most valuable during the concept, design, and construction phases of the project.

5. *The answers are:* **(A)** Administration
   (B) Materials In

The functional characteristic groups established under the EBO&M rating system are Materials In (which includes credits addressing the sustainable purchasing policy of a building), Materials Out, Administration (which includes credits addressing the planning and logistics support of operating a high-performance building), Green Cleaning, Site Management, Occupational Health and Productivity, Energy Metrics, and Operational Effectiveness. Sustainable Sites and Water Efficiency are credit categories, not functional characteristic groups. Waste Management is neither a credit category nor a functional characteristic group.

6. *The answers are:* **(B)** adhering to the plan throughout project
   (C) aligning goals with budget
   (E) establishing project goals and expectations

A project is more likely to stay within budget when its goals and budget are coordinated and when the project team adheres to the plan and frequently checks expenses against the budget. USGBC does not provide guidance for project budgeting. Submitting documentation for fewer credits does not necessarily lead to a successful LEED project budget.

7. *The answers are:* **(A)** accessible roof decks
   (C) non-vehicular, pedestrian-oriented hardscapes

Projects in urban areas often have little or no setback, but they can incorporate pedestrian hardscapes, pocket parks, accessible roof decks, plazas, and courtyards to meet the open space requirements.

On-site photovoltaics contribute to on-site renewable energy generation. Pervious parking lots contribute to on-site stormwater mitigation. Landscaping with indigenous plants may reduce the amount of water needed for irrigation. None of these strategies contributes to meeting open space requirements.

8. *The answers are:* **(A)** building orientation

    (B) envelope thermal efficiency

    (C) HVAC system size

A building's overall energy consumption can vary depending on building orientation, building location, envelope insulation, fenestration *U*-values, size of the HVAC system, and electricity requirements of lighting and appliances. Addressing these factors during the design can result in reduced energy bills for the life of the building. Refrigerant selection affects the building's environmental impact (but not the HVAC system efficiency), and volatile organic compound (VOC) content affects indoor air quality.

9. *The answers are:* **(C)** maintenance

    (E) utilities

Life-cycle costing is an accounting method used to evaluate the economic performance of a product or system over its useful life. Life-cycle cost calculations include maintenance and operating costs (including the cost of utilities). Neither the occupants' individual expenses nor the initial cost of an investment factor into a building's life-cycle costs.

10. *The answers are:* **(C)** reduced heat island effect

    (D) reduced light pollution

The primary purpose of increasing a building's daylighting is to reduce the need for electric light. Reducing electric light use reduces energy costs and the carbon dioxide emissions (or air pollution) created by the building. Additionally, statistics show that productivity is markedly better in daylit buildings than in buildings that rely heavily on electric lighting.

Light pollution is related to the amount of light transmitted from a building after hours. Daylighting is related to daytime lighting of the interior of the building. A building's heat islands are not affected by daylighting.

11. *The answers are:* **(B)** renovation of part of an owner-occupied building

    (C) tenant infill of an existing building

The LEED for Commercial Interiors (CI) rating system provides the opportunity for tenant spaces and parts of buildings to achieve LEED certification. Major building envelope renovations would be certified under the LEED for New Construction (NC) rating system. Upgrades to the operations and maintenance of an existing facility would be certified under the LEED for Existing Buildings: Operations & Maintenance (EBO&M) rating system.

Practice Exam Part One Answers

12. *The answers are:* (C) responsible party
    (D) time period

Each policy should identify the individual or team responsible for its implementation. Additionally, the time period over which the policy is applicable should be indentified. The time period is not necessarily the same amount of time as the performance period. Performance periods apply only to LEED EBO&M projects.

13. *The answer is:* (B) LEED rating systems

Chain-of-custody is the documented status of a product from the point of harvest or extraction to the ultimate consumer end use; recording the chain-of-custody can promote sustainable construction. Standard operating procedures (SOPs) are detailed written instructions that document a method with the intention of achieving uniformity. A waste reduction program helps a project team minimize waste by using source reduction, reuse, and recycling. Both SOPs and waste reduction programs promote sustainable operations. Of the answer options, only the LEED rating systems support design, construction, and operations.

14. *The answers are:* (C) primary contact information
    (D) project owner information
    (E) project type

To register a project for LEED certification, the registrant must provide account login information, primary contact information, project owner information, general project information, payment information, and the project type. Project team member names do not have to be submitted to complete the registration form. A LEED project is not required to have a LEED AP on its team.

15. *The answers are:* (A) achieve certification as a LEED home
    (E) market and sell the home

The five basic steps of LEED for Homes are

1. contact a LEED for Homes certification provider and join the program
2. identify the project team
3. build the home to the stated goals
4. achieve certification as a LEED home
5. market and sell the home

Becoming a USGBC member company reduces the registration fee; however, it is not a basic step.

16. *The answers are:* (D) hydrofluorocarbons
    (E) water

Using halons, chlorofluorocarbons, and hydrochlorofluorocarbons in fire suppression systems can lead to ozone depletion. Using water and hydrofluorocarbons will have a minimal effect on ozone depletion.

17. *The answer is:* **(D)** total area of the site including constructed and non-constructed areas

The total area within the legal property boundaries of the site is considered the property area. The project site area that includes constructed and non-constructed areas is the development footprint.

18. *The answer is:* **(C)** LEED EBO&M

LEED EBO&M is the only rating system that includes performance periods and a recertification requirement. LEED EBO&M projects may recertify as often as every year, and must recertify at least every five years to maintain their LEED certification status.

19. *The answer is:* **(B)** being pursued and some online documentation has been uploaded

A white check mark indicates that the credit is being pursued and has been assigned to a project member. Some documentation has been uploaded; however, additional info needs to be submitted.

20. *The answers are:* **(B)** energy use reduction
    (D) sustainable site use
    (E) water use reduction

Remodeled projects typically achieve Building Reuse credits within the Materials and Resources credit category; however, they may have difficulty achieving credits in the Sustainable Sites, Water Efficiency, and Energy and Atmosphere categories. This is because when the site is predetermined, the project team does not have the opportunity to select a more favorable site that would help in achieving those credits. Compared to newer buildings, older buildings usually have less energy-efficient insulation systems and less efficient plumbing fixtures.

21. *The answers are:* **(B)** interior finishes
    (C) lighting
    (D) mechanical distribution

The LEED Core & Shell (CS) rating system is designed to help designers, builders, developers, and new building owners increase the sustainability of a new building's core and shell construction. It covers base building elements and complements the LEED for Commercial Interiors (CI) rating system. Interior space layout, interior finishes, lighting, and mechanical distribution may not be directly controlled by the developer, and therefore if a project team wishes to include these elements in a project's LEED certification, LEED CS may not be appropriate. Site selection and the building's envelope insulation can be directly controlled by the owner when pursuing LEED CS.

22. *The answers are:* **(D)** project administrators to manage LEED projects
    (E) project team members to manage LEED prerequisites and credits

LEED Online does not provide advertising of any sort. Project administrators assign access responsibilities for prerequisites and credits to project team members using LEED Online. Local code officials are unable to view online LEED documentation. The Energy Star Target Finder tool will help a project team analyze anticipated building energy performance.

23. *The answers are:* (A) Heat Island Effect
    (E) Stormwater Design

Green roofs help mitigate stormwater and reduce the roof's heat island effect by increasing evapotranspiration (which has a cooling effect), and increasing the roof's albedo. Light pollution reduction is achieved through strategic lighting design. Site selection credit is achieved by not locating the building, or hardscapes, on a prohibited site.

24. *The answer is:* (D) reduced toxic waste

Because mercury waste is toxic, using lightbulbs with low mercury content, long life, and high lumen output will result in reduced toxic waste. The mercury content of lights does not affect light pollution, which is the impact of artificial light on night sky visibility.

25. *The answers are:* (A) light to night sky
    (B) light trespass
    (D) stormwater mitigation

Site lighting and stormwater mitigation must be addressed when designing a sustainable site. On-site renewable energy can reduce the building's burden on the power grid and is addressed in the Energy and Atmosphere category. Refrigerants affect ozone depletion and are addressed in the Energy and Atmosphere category.

While important, controlling light trespass and light to night sky are not prerequisites for sustainable site design.

26. *The answers are:* (B) evaluate durability risks of project
    (C) incorporate durability strategies into design

The four basic elements of a durability plan are evaluation of durability risks, incorporation of durability strategies into design, implementation of durability strategies into construction, and completion of a third-party inspection of the implemented durability features.

27. *The answers are:* (A) ensure the technical soundness of the LEED reference guides and training
    (C) resolve issues to maintain consistency across different LEED rating systems

Technical Advisory Groups (TAGs) respond to Credit Interpretation Requests (CIRs) and assist in the development of LEED credits. The Technical Scientific Advisory Committee (TSAC) ensures LEED and its supporting documentation is technically sound while assisting USGBC with complex technical issues.

28. *The answers are:* (B) preindustrial material
    (C) rapidly renewable material
    (D) regionally extracted material

The LEED EBO&M rating system defines the amount of material that must come from post-consumer, pre-industrial, rapidly renewable, or regionally extracted sources in order to earn credit for ongoing consumable purchases. The Rainforest Alliance certifies food, and is not related to ongoing consumables. Salvaged material use and purchase can contribute to earning credit for the sustainable purchases of durable goods and facility alterations, but not for ongoing consumables.

29. *The answer is:* (A) LEED CI

LEED for Commercial Interiors (CI) addresses tenant spaces within a building. Both LEED for New Construction & Major Renovation (NC) and LEED for Existing Buildings: Operations & Maintenance (EBO&M) are rating systems that apply to entire buildings. LEED for Core & Shell (CS) addresses buildings that are built with no, or limited, interior buildouts.

30. *The answer is:* (D) geographic location

USGBC chapters and regional councils identify which credits are eligible for Regional Priority points based on the needs of each environmental zone. LEED Online determines the region of a project based on its geographic location, which it identifies from the project site's zip code.

31. *The answers are:* (B) construction cost
    (C) documentation cost
    (E) soft cost

Project teams should consider the potential construction, soft, and documentation costs before committing to pursing a particular LEED credit. The application review cost is established regardless of the number or selection of LEED credits pursued, and is based on the building's floor area. Registration cost is the same for every LEED project, varying only depending on if the project is registered by a member or non-member company.

32. *The answer is:* (C) LEED EBO&M

Credit Interpretation Requests (CIRs) and appeals are components of every LEED rating system. LEED EBO&M is the only rating system that requires the implementation of performance periods.

33. *The answer is:* (D) ANSI/ASHRAE/IESNA 90.1

ANSI/ASHRAE/IESNA 90.1 sets minimum requirements for the energy-efficient design of all buildings except low-rise residential buildings. ANSI/ASHRAE 52.2 addresses air cleaner efficiencies; ASHRAE 55 addresses thermal comfort; and ANSI/ASHRAE 62.1 addresses ventilation.

Practice Exam Part One Answers

34. *The answers are:* (A) Controllability of Systems
(C) Indoor Chemical and Pollutant Source Control

Project teams intending to increase a building's ventilation will have to consider the implications for the building's commissioning, measurement and verification, and energy performance, all of which will be directly affected.

Controllability of Systems relates more to the thermal comfort and lighting of a building than the building's ventilation. Increasing ventilation will not help or prevent a project team from achieving Indoor Chemical and Pollutant Source Control.

35. *The answer is:* (B) to track the movement of wood products from the forest to the building

A chain-of-custody document verifies compliance with Forest Stewardship Council (FSC) guidelines for wood products, which requires documentation of every movement of wood products from the forest to the building.

36. *The answers are:* (D) manufactured within 500 miles of the project site
(E) permanently installed on the project site

For the purposes of the LEED rating system, regional materials are permanently installed building components that have been extracted, harvested, recovered, or manufactured within 500 miles of the project site.

Regional materials do not need to be recycled or post-consumer materials, nor do they need to be rapidly renewable (agricultural products that take 10 years or less to grow or raise and can be harvested in an ongoing and sustainable fashion). Forest Stewardship Council (FSC) certification applies only to wood and is not a requirement of regional materials.

37. *The answer is:* (C) Materials Reuse

A project team's choice of paint will have little or no effect on materials reuse. Choosing white exterior paint can contribute to reducing a building's heat island effect. Choosing paint with low volatile organic compounds (VOCs) will contribute to earning Low-Emitting Materials credit. Choosing bio-based paint can help a project team earn Rapidly Renewable Materials credit.

38. *The answers are:* (A) core credits
(C) innovation credits
(E) prerequisites

All LEED rating systems contain prerequisites, core credits, and innovation credits. Sustainable operations and educational programs may help a project team achieve either a core credit or an innovation credit.

39. *The answer is:* (B) door

Computers, office desks, and landscaping equipment are examples of durable goods, which are defined by the LEED reference guides as goods with a useful life of two years or more and that are replaced infrequently. Doors are considered part of the base building equipment.

40. *The answers are:* (C) local climate

                      (D) operational costs

Installing a larger ventilation system will minimally impact construction costs, but will significantly increase the energy cost throughout the life cycle of the building. Prior to increasing the ventilation rate of a building, the design winter and summer temperatures and humidity should be considered. Regardless of building location, the interior temperature should typically be between 68°F and 74°F and the relative humidity should be between 50% and 55%. Refrigerant management is not affected by ventilation systems.

41. *The answers are:* (B) global warming potential

                      (C) ozone depletion potential

Refrigerants are chemical compounds that, when released to the atmosphere, deteriorate the ozone layer and increase greenhouse gas levels. LEED requires project teams to consider the ozone depletion potential and the global warming potential of refrigerants used in a building's HVAC & R system.

42. *The answers are:* (B) email address

                      (C) industry sector

                      (E) phone number

There is no cost to set up a personal user account on the USGBC website. The individual does not need to be a USGBC member, have prior LEED project experience, have or intend to have LEED credential, or supply his or her company name.

43. *The answers are:* (B) CIRs reviewed by a TAG for the team's own project

                      (D) CIRs posted prior to project application, for the applicable rating system only

Project teams are required to adhere only to CIRs uploaded to the USGBC website prior to project registration. Adherence is required for those submitted after project registration only if they were submitted by the project team itself. CIR requirements must be adhered to regardless of the project's geographic location; however, project teams are generally only required to follow CIRs for their specific rating system.

44. *The answers are:* (C) achieve the next incremental level of an existing credit

                      (D) double the requirements of an existing credit

Innovation in Design points for exemplary performance are earned for going above and beyond existing credit requirements. Alternatively, project teams can earn ID points for achieving the next incremental level of an existing credit if it is specified within the corresponding rating system.

45. *The answers are:* (B) collecting rainwater for sewage conveyance

                      (D) using cooling condensate for cooling tower make-up

As with every sustainable strategy, consult all applicable codes prior to implementation. Non-potable water used for cooling tower makeup or sewage conveyance can lead to earning LEED credit. On-site septic tanks do not reduce potable water used for sewage conveyance, and therefore do not contribute to the achievement of a LEED point for water conservation. Code requirements restrict project teams from using blackwater for landscape irrigation.

46. The answer is: **(C)** LEED credit equivalence submittal

Project teams intending to achieve Innovation in Design credit for innovation (not exemplary performance) must follow the LEED credit equivalence process, which requires the following.

- the proposed innovation credit intent
- the proposed credit requirement for compliance
- the proposed submittal to demonstrate compliance
- a summary of potential design approaches that may be used to meet the requirements

The Credit Interpretation Requests (CIRs) process does not involve the proposition of new credits, and therefore does not require an explanation of a proposed credit requirement. The LEED Online submittal templates do not allow for the proposal of new credits.

47. The answer is: **(B)** Center for Resource Solutions

The Center for Resource Solutions' Green-e energy program is a voluntary certification and verification program for renewable energy products.

Energy Star's Portfolio Manager is a federal program that helps businesses and individuals protect the environment through energy efficiency. The Department of Energy's mission is to advance the energy security of the United States. There is no such thing as the Center for Research and Development of Green Power.

48. The answers are: **(B)** International Energy Conservation Code (IECC)
 (D) International Mechanical Code (IMC)
 (E) International Plumbing Code (IPC)

The International Code Council (ICC) is a consolidated organization that comprises what was formerly the Building Officials and Code Administrators International, Inc. (BOCA), the International Conference of Building Officials (ICBO), and the Southern Building Code Congress International, Inc. (SBCCI). The ICC family of codes includes, but is not limited to, the International Building Code (IBC), the International Fire Code (IFC), the International Plumbing Code (IPC), the International Mechanical Code (IMC), and the International Energy Conservation Code (IECC).

49. The answer is: **(C)** Companies can be USGBC members.

People can be LEED accredited, buildings can be LEED certified, and companies (not individuals) can be USGBC members. It is not a requirement that a LEED project administrator (or anyone else involved with a LEED project) be a LEED AP.

50. *The answer is:* **(B)** 80%

A project team earning more than 40% but less than 50% of core credits within the LEED rating systems will earn a LEED Certified plaque. They will earn a LEED Silver plaque for earning more than 50% but less than 60% of Core Credits; LEED Gold for earning more than 60% but less than 80%; and LEED Platinum for earning more than 80%.

51. *The answers are:* **(C)** a means to submit a project for review

(E) the project's final scorecard

Project teams submit their projects to the Green Building Certification Institute (GBCI) for review using the LEED submittal templates, which also generate the final scorecard for LEED projects.

Project team members can manipulate the documentation for only those prerequisites or credits assigned to them. The LEED reference guides contain potential strategies for credit and prerequisite approval. Credit Interpretation Requests are reviewed and submitted at LEED Online, but not though submittal templates.

52. *The answers are:* **(A)** bid

(B) design

(D) pre-design

According to the *Sustainable Building Technical Manual*, an environmentally responsive design process includes the following key steps: pre-design, design, bid, construction, and occupancy. While the selection of vendors, consultants, and/or contractors is part of the process, it is not a key step. Rebid and post-design should not occur if an appropriate process has been followed.

53. *The answer is:* **(C)** Projects can earn half points under the Site Selection credit.

LEED EB and LEED EBO&M projects must recertify at least every five years to maintain their LEED certification, but the LEED CI rating system does not have a recertification option. Precertification is only available if a project team is utilizing the LEED for Core & Shell rating system. Prohibiting smoking is an option in every LEED rating system for meeting the Environmental Tobacco Smoke prerequisite; because this is a prerequisite, no points are awarded for compliance.

54. *The answers are:* **(B)** clarify

(D) denied

(E) earned

An *earned* ruling requires no additional action from the project team. A *clarify* ruling requires the project team member to address the issues of the project reviewer. A *denied* ruling indicates that the project team member either misunderstood the intent and/or failed to meet the prerequisite or credit. *Anticipated* and *deferred* are categories for prerequisites and credits reviewed during the design phase submittal.

55. *The answers are:* (A) delegate responsibility and oversee all LEED committee activities
(D) establish and enforce LEED direction and policy

The role of the LEED Steering Committee is to establish and enforce LEED direction and policy as well as to delegate responsibility and oversee all LEED committee activities.

Technical Advisory Groups (TAGs) respond to Credit Interpretation Requests and the GBCI develops and administers the accreditation exams. The Technical Scientific Advisory Committee (TSAC) ensures LEED and its supporting documentation is technically sound.

56. *The answers are:* (B) assign credits to project team members
(C) build a project team
(E) submit projects for review

Project administrators can do many things through the LEED Online workspace, including assigning credits to project team members, building the project team, and submitting projects for review. Previously submitted Credit Interpretations Requests (CIRs) can be viewed by USGBC company members as well as LEED project team members; however, credit appeals submitted by other project teams are not available for review. Precertification is a unique aspect of the LEED for Core & Shell rating system.

57. *The answers are:* (E) LEED for Homes rating system
(F) sample LEED submittal templates
(G) usgbc.org account

A free download of the LEED for Homes rating system is available at www.usgbc.org/homes. Sample LEED submittal templates are available at LEED Online. An individual may create a free usgbc.org account.

Each Credit Interpretation Request (CIR) requires that a fee be submitted to USGBC before a Technical Advisory Committee (TAG) will review it. The *LEED Reference Guide for Building Design and Construction* is available for purchase at **www.ppi2pass.com/LEED**. USGBC and LEED brochures are available for purchase at www.gbci.org/publications. LEED certification requires project registration, whose fees are described at www.usgbc.org/leedregistration.

58. *The answers are:* (A) additional response time may be needed
(C) it is sent to the LEED Steering Committee

CIRs beyond the expertise of the Technical Advisory Group (TAG) are sent to the LEED Steering Committee and/or relevant LEED Committees for a ruling. Additional time is typically needed in this situation.

59. *The answer is:* (A) LEED project tools

LEED project tools are only available to registered project teams. A directory of all LEED registered and certified projects, as well as posted rulings on Credit Interpretation Requests are available to registered USGBC members. LEED Online does not contain a list of LEED project credit appeals. While the project's submittal templates also become available after the project is registered, sample submittal templates are available to the public.

60. *The answers are:* (A) codes and regulations that address asbestos and water discharge
(B) codes and regulations that address PCBs and water management

Buildings must be in compliance with federal, state, and local environmental laws and regulations, including but not limited to those addressing asbestos, PCBs, water discharge, and water management. LEED certification can be revoked upon knowledge of noncompliance.

Complying with referenced standards that address sustainable forest management practices, fixture performance requirements for water use, and volatile organic compounds (VOCs) may help achieve LEED credits; however, doing so is not a program requirement.

61. *The answers are:* (A) include members from varying industry sectors
(E) involve the team in different project phases

An integrated project team actively involves participants from varying industry sectors throughout the project, and meets monthly to review the project status.

The role of the commissioning authority is to verify the mechanical systems are operating as the designer intended. Product vendors may provide valuable insight that promotes the success of the project; however, they are not required on the project team. The purpose of including a LEED AP on a project team is to encourage LEED design integration and to streamline the application and certification processes, but including a LEED AP is not needed to create an integrated project team.

62. *The answer is:* (C) LEED EBO&M

Facilities undergoing minor envelope, interior, or mechanical changes must pursue LEED certification under the EBO&M rating system. Facilities undergoing major envelope, interior, or mechanical changes must pursue LEED certification under the NC rating system. LEED CI is used to certify tenant spaces while LEED CS is used to certify buildings prior to tenant infill.

63. *The answers are:* (D) use rainwater for sewage conveyance or landscape irrigation
(E) remediate contaminated soil

Soil remediation may help a project team achieve a credit under Brownfield Redevelopment, and using collected rainwater may earn a team points under the Water Efficiency category.

Erosion and sedimentation control, installing HCFC-based HVAC & R equipment, and the storage and collection of recyclables are LEED certification prerequisites (not credits).

64. *The answer is:* (D) To achieve LEED certification, every prerequisite and a minimum number of points must be earned.

The first step toward LEED certification is project registration. A fee must be paid when submitting Credit Interpretation Requests (CIRs). There is a LEED rating system for every type of building, including non-commercial buildings.

65. *The answer is:* (B) Optimize Energy Performance

Ground source heat pumps are energy-efficient mechanical systems that may help project teams earn the energy performance prerequisite and credit. A ground source heat pump makes use of a vapor compression refrigeration cycle, which requires electricity to operate, and therefore it is not a renewable source of energy. These pumps do not generate energy (so they don't contribute to On-Site Renewable Energy credits), nor do they reduce the amount of water used by a project.

66. *The answers are:* **(D)** offer a volume certification path
　　　　　　　　　　 (E) provide a streamlined certification process for large-scale projects

The goals of the Portfolio Program are to assist participants in integrating green building design, construction, and operations into their standard business practices using LEED technical standards and guidance; provide a cost-effective, streamlined certification path for multiple buildings that are nearly identical in design; recognize leaders who are creating market transformation through their commitments and achievements in green building; foster a network of investors, developers, owners, and managers committed to systemically greening their building portfolios; and support participating organizations in fulfilling their sustainability commitments by providing solid performance metrics that can be given to stakeholders.

Submittal templates provide a template of key data for the design team members to compile. The mission of LEED in general is to encourage and accelerate the "global adoption of sustainable green building and development practices through the creation and implementation of universally understood and accepted standards, tools, and performance criteria."

67. *The answers are:* **(A)** before the design phase
　　　　　　　　　　 (B) during the construction phase

The LEED project budget should be addressed before the design phase and throughout the construction phase of the project. LEED project team selection is based on the created budget.

68. *The answer is:* **(C)** installing native or adapted plants

Turf grass typically requires sizable amounts of water to sustain it, while native or adapted plants can survive on the amount of rainfall a site naturally receives. Projects with no landscaping are not eligible to earn landscaping credits. Invasive plants such as weeds are not considered landscaping by the LEED reference guides.

69. *The answer is:* **(B)** 4 points

Regionalization is an opportunity for project teams to earn additional points in the Innovation in Design category of each rating system. GBCI determines which six credits are priorities in each region of the United States. Project teams in each region can earn up to four additional points for achieving the Regional Priority credits assigned to its region.

70. *The answer is:* **(B)** Energy Metrics

Energy Metrics credits focus on the building's energy performance and ozone protection.

Energy and Atmosphere is a credit category within the LEED rating systems, and is not a functional characteristic grouping. Materials Out credits are associated with the sustainable solid waste management policy of a building. Site Management credits address sustainable landscape management practices.

71. *The answers are:* (B) interior moisture loads

(D) pests

(E) ultraviolet radiation

The intent of durability planning is to appropriately design and construct high performance buildings that will continue to perform well over time. The principle risks are exterior water, interior moisture, air infiltration, interstitial condensation, heat loss, ultraviolet radiation, pests, and natural disasters.

72. *The answer is:* (B) contact a LEED for Homes provider

The first step for participating in the LEED for Homes program is to contact a LEED for Homes provider; only then is the project registered with GBCI. Becoming a LEED AP will strengthen your LEED knowledge; however, it is not required to pursue LEED certification in any rating system. Credit Interpretation Requests can be submitted only after a project is registered.

73. *The answer is:* (D) refrigerant leakage

HVAC & R equipment with a relatively short equipment life, a relatively high refrigerant charge, and/or refrigerant leakage will contribute to ozone depletion and global warming.

74. *The answer is:* (B) durability plan

As explained in the LEED for Homes rating system, a successful durability plan includes both a ranking of durability risks and strategies to minimize the risks, which could include sealing ventilation ducts, installing rodent- and corrosion-proof screens, and using air-sealing pump covers.

The PE exemption form gives project teams the opportunity to follow a streamlined path to achieve certain prerequisites and credits. A building's landscape management plan focuses on ecology and wildlife outside of the building.

75. *The answer is:* (A) ASHRAE 55

ASHRAE 55 was created to establish acceptable indoor thermal environmental conditions. ANSI/ASHRAE 52.2 addresses air cleaner efficiencies; ANSI/ASHRAE 62.1 addresses ventilation; ANSI/ASHRAE/IESNA 90.1 addresses building efficiency.

76. *The answers are:* (A) email address

(D) organization name

(E) individual's title

The primary contact of a LEED project is not required to be a LEED AP or have previous LEED project experience.

Practice Exam Part One Answers

77. *The answers are:* (D) project summary
(E) project team members

The project team and project summary must be confirmed prior to the LEED application process. CIRs and precertification are items available for LEED project administrators; however, they are not required items. The project cost does not need to be confirmed prior to registration.

78. *The answer is:* (A) evapotranspiration

Evapotranspiration is the term used to describe water that is lost through plant transpiration and evaporation from soil.

79. *The answer is:* (D) view a protected electronic version of the LEED reference guide

An online reference guide access code is provided with the purchase of a LEED reference guide. The code provides its owner with electronic access to the reference guide. Project team members can also acquire the access code for 30 days of electronic access to the reference guide from a project team administrator after joining a LEED project.

80. *The answer is:* (C) intent

The structure of LEED prerequisites and credits is considered part of the LEED brand and includes intent, requirements, and potential technologies and strategies. This structure is maintained in each version of the LEED rating system. Economic and environmental benefits, greening opportunities, and submittal requirements are included in the rating systems and/or LEED online.

81. *The answers are:* (B) durable goods
(E) waste stream

The environmental impact of refrigerants and energy consumption are addressed in the Energy and Atmosphere category. Materials and Resources credits address the flow of materials to and from a project site as well as the waste stream generated from a project building.

82. *The answer is:* (D) urban area

Building on a previously developed site helps conserve existing greenfields (undeveloped land). Locating a project in an urban location will provide the occupants with increased opportunities to utilize public transportation and reduce the need to drive to local amenities. LEED prohibits projects from building on public parkland, due to the potential economic and environmental implications.

83. *The answer is:* (A) declarant's name

Every LEED submittal template begins with the following statement: "I, *declarant's name*, from *company name* verify that the information provided below is accurate, to the best of my knowledge."

Product manufacturer info may be required for some documentation; however, it is not a requirement of every submittal. The project location is identified in the LEED registration form.

84. *The answers are:* (B) establishing a project budget
    (C) establishing project goals
    (D) site selection

The pre-design phase of every LEED project includes developing a green vision, project goals, priorities, building program, and budget; assembling a green team; developing partner strategies and project schedules; researching and reviewing local codes, laws, standards; and selecting a project site. Commissioning, testing and balancing, and training are steps of the construction phase.

85. *The answer is:* (A) A CIR must be submitted as a text-based inquiry.

There is no mechanism available to submit attachments during the CIR process. A complete project narrative is not required, and the CIR is limited to 600 words.

86. *The answers are:* (D) energy efficiency
    (E) environmental impact
    (F) indoor environment

Green building design and construction should be guided primarily by energy efficiency, environmental impact, resource conservation and recycling, indoor environmental quality, and community considerations. Design and bid costs should be considered; however, green building focuses primarily on life-cycle costs. Construction documents are created once the guideline issues have been addressed.

87. *The answers are:* (A) accreditation of industry professionals
    (C) certification of sustainable buildings

GBCI administers the LEED accreditation for industry professionals, and is responsible for the certification of buildings and parts of buildings. It does not certify products. USGBC provides sustainable education programs.

88. *The answers are:* (A) appeal documentation
    (D) construction documentation

Credits can be earned during the construction phase of a project. Design credits can be submitted in the design review, but they will only be distinguished as "anticipated" or "denied." Project teams can modify a strategy and resubmit a design submittal credit in the construction phase to change the designation from "denied" to "earned". The construction review must include all design and construction credits that are pursued by the project team. After the construction review, credits are designated as "earned" or "denied." This designation can only be changed by submitting an appeal and paying a fee for each credit appealed. After the credit is appealed, it will be designated "earned" or "denied." The credit cannot be appealed a second time.

Practice Exam Part One Answers

89. *The answers are:* **(D)** promotes design efficiency

(E) reduces design and construction time

Commissioning increases a project's initial cost, but should reduce its life-cycle cost. Commissioning authorities are not responsible for ensuring a project's compliance with a code. Rather, they verify that review design documents to help eliminate changes during construction or contractors from making design decisions. The commissioning agent's check helps prevent consultant redesign, as well as on-the-job engineering from the contractors, and therefore reduces the overall time spent on design and construction.

90. *The answers are:* **(A)** anthropogenic nitrogen oxide

(B) carbon dioxide

(D) sulfur dioxide

Conventional fossil-fuel generated electricity (such as that from coal-fueled power plants and liquid petroleum) results in the release of carbon dioxide into the atmosphere, which contributes to global warming. Coal-fired electric utilities emit both anthropogenic nitrogen oxide (nitrogen oxide that is a result of human activities, and which is a key contributor to smog) and sulfur dioxide (a key contributor to acid rain).

91. *The answer is:* **(D)** shading hardscapes with vegetation

Heat island effect can be minimized by having high (not low) solar reflectance index (SRI) values, minimizing the area of hardscapes, and shading necessary hardscapes with trees and bushes. The glazing factor is related to daylighting, not heat island effect.

92. *The answers are:* **(B)** indoor environmental quality criteria

(C) mechanical system descriptions

(E) references to applicable codes

Mechanical systems (which include HVAC & R, plumbing, and electrical systems) must be addressed in the basis of design (BOD). The BOD must also establish the procedure for the installed mechanical equipment to achieve the required indoor environmental quality criteria. Applicable codes must also be included to help provide guidance to the installing contractors. Process equipment and building materials may be addressed here; however, it is not a requirement.

93. *The answers are:* **(B)** energy sustainability consultant

(C) landscape architect

Utility managers, product manufacturers, and code officials may be a good resource for sustainability advice; however, they are not directly involved with the decisions of the project, and therefore are not considered part of the integrated project team.

94. *The answer is:* **(A)** 50%

Replacement or upgrade to 50% of the building's envelope (walls, floors, and roof) is considered a major renovation. Replacement or upgrade to 50% of the building's interior (nonstructural walls, floor coverings, and drop ceilings) is considered major. Replacement or upgrade to 50% of the building's mechanical systems (HVAC & R, lighting, plumbing) is considered major. The percentages are calculated using either the cost of the renovation or the area being renovated.

95. *The answer is:* **(B)** 50%

Owners can occupy up to 50% of the building's leasable space and still be eligible to pursue LEED certification under the Core & Shell rating system. Buildings in which the owner occupies more than 50% of the leasable floor space must pursue LEED certification under the New Construction rating system.

96. *The answers are:* **(A)** Energy and Atmosphere
(F) Operational Effectiveness

Credits and prerequisites can be grouped by the credit categories designated within the LEED rating systems, or by functional characteristics. The Best Management Practices prerequisite falls within in the Energy and Atmosphere LEED credit category, but can also be grouped as an Operational Effectiveness prerequisite by its functional characteristic. Operational Effectiveness credits and prerequisites support best management practices.

97. *The answers are:* **(B)** gross area of the building
(D) project budget
(E) site conditions

The project's primary contact is the only individual required to submit their company's name. The project team creates a list of possible innovation strategies, which is submitted during the project application phase.

98. *The answer is:* **(B)** previously certified under the LEED EBO&M rating system

LEED for Existing Buildings: Operations & Maintenance is the only rating system that has a recertification option available. Precertification does not exist for the LEED for Schools rating system.

99. *The answers are:* **(C)** installing heat recovery systems
(E) zoning mechanical systems

Increasing the ventilation rate and performing a flush out before occupancy improves indoor air quality but increases the amount of energy used. Reducing a building's heat island effect will not affect the building's energy use.

100. *The answer is:* **(D)** VOC content in building materials

A building's indoor environmental quality can be improved by controlling noise pollution, providing as much natural lighting as possible, and providing adequate ventilation (thereby improving the air quality). Volatile organic compounds (VOCs), which damage lung tissue, should be minimized.

Practice Exam Part Two Answers

Answers begin on the page that follows.

LEED Homes Practice Exam

#	Ans	#	Ans	#	Ans
1.	A, B	35.	C	68.	D
2.	B	36.	B, C	69.	D
3.	B, C, E	37.	B, D	70.	A
4.	D	38.	A	71.	B
5.	C, D	39.	A, C	72.	C
6.	A, D, E	40.	A	73.	A
7.	B, E	41.	B, D	74.	C, D
8.	A, C	42.	A, B	75.	A
9.	C	43.	B	76.	D
10.	C, D	44.	A, B, D	77.	B
11.	B	45.	B	78.	C, E, F
12.	D	46.	C	79.	A
13.	C	47.	A	80.	D
14.	D, E	48.	B, C, D	81.	A, B
15.	A, B, C	49.	B	82.	A, C, D
16.	B, C	50.	A, B	83.	C
17.	D	51.	C	84.	A, C, D
18.	B	52.	A	85.	B
19.	C	53.	A	86.	D
20.	D	54.	C	87.	C
21.	A	55.	A, B, C	88.	B
22.	A, C, D	56.	A, C, F	89.	B
23.	B, D, E	57.	C	90.	B, F
24.	A, B	58.	B	91.	A, B, C
25.	B	59.	B	92.	D
26.	A, C, D	60.	D	93.	B
27.	C	61.	B, C, F	94.	C, D, E
28.	B	62.	C	95.	C
29.	A	63.	B	96.	D
30.	C	64.	A	97.	B, D
31.	C	65.	A, B, E	98.	A, C, D, F
32.	C, D	66.	B, D	99.	B, E
33.	D, E	67.	C	100.	B
34.	B				

Practice Exam Part Two Answers

1. *The answers are:* **(A)** discussion of furniture and appliance selections

    (B) explanation of the LEED Scorecard

AE Prerequisite 1.1, Education of the Homeowner or Tenant: Basic Operations Training, describes two requirements for appropriate occupant education. One requirement is a one-hour walk-through, during which the occupant is shown all installed equipment and is educated on proper equipment operation and maintenance. Although discussion of the LEED scorecard and appropriate furniture and appliance selection are part of the education of the homeowner or tenant, they are not required during the one-hour walk-through.

2. *The answer is:* **(B)** 10 points

The Location and Linkages (LL) category provides two pathways for earning credits. The first pathway is to meet the requirements of LL Credit 1, LEED for Neighborhood Development, for which a project will earn 10 points. The second pathway encompasses LL Credits 2 through 6, which together are worth up to 10 points. A project that follows the first pathway is not eligible to earn points for the second pathway, and vice versa.

3. *The answers are:* **(B)** prime soils

    (C) unique soils

    (E) soils of state significance

LL Credit 2, Site Selection, specifies that it is not acceptable to build a LEED home on prime soils, unique soils, or soils of state significance when the site is previously undeveloped. All other soil types are acceptable. Alfisols and ustic soil are common soil types found throughout the U.S. in sub-humid and semi-arid regions. A project team can find a site's soil type by reviewing the soil surveys, maps, and data made available online through the USDA's Natural Resources Conservation Service.

4. *The answer is:* **(D)** 45%

A 20% to 44% reduction in estimated irrigation water usage will earn a project between 2 and 6 points for SS Credit 2.5, Landscaping: Reduce Overall Irrigation Demand by at least 20%, and 0 points for WE Credit 2.3, Irrigation System: Reduce Overall Irrigation Demand by at least 45%. A reduction of 45% or more will earn 6 points for SS Credit 2.5 and between 1 and 4 points for WE Credit 2.3. In other words, for a project to earn points for both WE Credit 2.3 and SS Credit 2.5, the minimum estimated irrigation water usage reduction is 45%.

5. *The answers are:* **(C)** kitchen

    (D) laundry room

Indoor moisture control measures are outlined in ID Credit 2, Durability Management Process, which, among other things, requires that water-resistant flooring be installed in all kitchens, bathrooms, laundry rooms, spa areas, and entryways.

6. *The answers are:* (A) duct leakage

    (D) local exhaust

    (E) outdoor air flow

It is the green rater's responsibility to conduct certain field inspections and performance tests of the completed home. Required green rater testing includes that of envelope leakage, duct leakage, outdoor air flow, local exhaust, and supply air flow. HVAC refrigerant charge testing is done by an HVAC professional. Indoor air quality is not specifically tested.

7. *The answers are:* (B) irrigation system inspection checklist

    (E) thermal bypass checklist

AE Prerequisite 1.1, Education of the Homeowner or Tenant: Basic Operations Training, requires that the operations and maintenance manual provided to the home's occupants include a copy of each of the following: the completed checklist of LEED for Homes features; all signed accountability forms; the durability inspection checklist; manufacturers' manuals for all installed equipment, fixtures, and appliances; general information on efficient use of resources; and general recommendations for operating and maintaining the home.

8. *The answers are:* (A) WE Credit 3, Indoor Water Use

    (C) EA Credit 7, Water Heating

Low-flow showerheads and faucets reduce demand for hot water, which decreases indoor water use and makes the project eligible to receive points under WE Credit 3.1, Indoor Water Use: High-Efficiency Fixtures and Fittings. Reducing demand for hot water may also significantly reduce the energy used for water heating, which is the objective of EA Credit 7, Water Heating.

9. *The answer is:* (C) 6 points

EA Credit 10, Renewable Energy, encourages installing renewable energy generation systems, which reduce a home's consumption of nonrenewable energy. For every 3% of a home's annual *reference* energy load that is met by the renewable system, a project can earn one point (up to 10 points). The *actual* electricity load is not used for this calculation. In the given scenario, the number of points is calculated as follows.

$$\text{number of points} = \frac{\frac{\text{annual renewable energy generated}}{\text{annual reference electricity load}}}{0.03} = \frac{\frac{2000 \text{ kW}}{10{,}000 \text{ kWh}}}{0.03}$$

$$= 6.6 \text{ points} \quad (6 \text{ points})$$

10. *The answers are:* (C) installing a central vacuum system

    (D) installing permanent walk-off mats

EQ Credit 8.2, Contaminant Control: Indoor Contaminant Control, describes appropriate measures for controlling indoor contaminants. These measures include installing permanent walk-off mats, central vacuum exhausting outside, and permanent shoe storage near the principal point of entry. Although flushing the home for 48 hours, having a detached garage, and sealing all ducts and vents during construction help reduce contaminants, they are not specified in this credit, but in EQ Credits 8.3, 10.4, and 8.1, respectively.

Practice Exam Part Two Answers

11. *The answer is:* **(B)** MR Credit 2, Environmentally Preferable Products

Because VOCs degrade air quality, it is preferable to use products with no or low VOC emissions. The acceptable VOC levels for products typically used in home construction are defined in MR Credit 2, Environmentally Preferable Products.

12. *The answer is:* **(D)** U.S. DOE

U.S. climate zones are defined and maintained by the U.S. DOE, which has also specified building envelope requirements for each climate zone. Knowing a project's climate zone helps a project team design for efficient energy, water, and material use. The U.S. DOE's prescriptive package information can help a project team accurately model energy performance, select plant species for landscaping, and design for regional climate considerations.

13. *The answer is:* **(C)** International Residential Code

One- and two-family dwellings of three stories or less are regulated by a stand-alone International Residential Code (IRC) that includes building, plumbing, mechanical, fuel gas, energy, and electrical provisions. The International Building Code (IBC) covers all buildings except detached one- and two-family dwellings and townhouses not more than three stories in height. The Building Officials and Code Administrators (BOCA) National Building Code is now part of the IBC, and there is no such thing as the Federal Building Code.

14. *The answers are:* **(D)** pine

(E) soapstone

The LEED definition of a rapidly renewable material or product is one that is grown and harvested within a 10 year life cycle. Agrifiber (which includes grass straw, cereal straw, rice straw, and similar materials) and cork both meet the definition of a rapidly renewable material. Linoleum is made of flax seed, which also falls in this category. Pine is fast-growing, relative to other hardwoods, but it is not typically harvested within 10 years. Soapstone, a rock largely composed of mineral talc, also is not a rapidly renewable material.

15. *The answers are:* **(A)** attic ceiling

(B) attic eave vents

(C) common walls between dwelling units

EA Credit 2, Insulation, refers to the Energy Star Qualified Homes Thermal Bypass Inspection Checklist, which identifies six areas of a home that must be inspected. The areas include overall air barrier and thermal barrier alignment; walls adjoining exterior walls or unconditioned spaces; floors between conditioned and exterior spaces; shafts; attic/ceiling interface; and common walls between dwelling units.

16. *The answers are:* **(B)** graywater system

(C) high-efficiency irrigation system

SS Credit 4, Surface Water Management, outlines multiple strategies to reduce a site's surface water runoff. Bioswales, permeable paving, and rain gardens all encourage the re-absorption of surface water, thus reducing the site's surface water runoff. Graywater systems and high-efficiency irrigation systems reduce the use of potable water, but they do not reduce water runoff.

17. *The answer is:* **(D)** 3 points

LL Credit 5, Community Resources/Transit, (worth up to 3 points) requires a home to meet criteria related to proximity and availability of community resources and transit services. To earn 3 points for the credit, the home must be located within a quarter mile of 11 basic community resources; within a half mile of 14 basic community resources; or within a half mile of transit services that offer 125 transit rides or more per weekday.

18. *The answer is:* **(B)** low-flow fittings

WE Credit 2, Irrigation System, presents multiple strategies for designing and installing a high-efficiency irrigation system. Installing low-flow fittings (such as faucets and showerheads) will decrease indoor water use, not irrigation water use. Moisture sensor controllers signal the irrigation system to replace only the moisture lost; pressure regulating devices prevent misting; and central shut-off valves provide further control over the system.

19. *The answer is:* **(C)** thermal conductivity

U-value (also called U-factor) is a measure of thermal conductivity (how much heat passes through an element) and is often used to rate the insulating quality of windows. A window with a low U-value is more energy efficient than a window with a higher U-value. The R-value, the inverse of U-value, is a measure of thermal resistance (an element's insulating ability).

20. *The answer is:* **(D)** U.S. EPA

The U.S. EPA has developed a radon risk map that divides the U.S. into three radon risk zones. The U.S. EPA encourages action when radon levels are above 4 picocuries per liter, which is most common in the map's zone 1. If a LEED for Homes project is located in zone 1, the homeowner must have a radon mitigation system installed, as defined by EQ Prerequisite 9.1, Radon Protection: Radon-Resistant Construction in High-Risk Areas. If a project is located in zones 2 or 3 and the homeowner decides to have a radon mitigation system installed, the project can earn 1 point.

21. *The answer is:* **(A)** MERV 8

Each air filter with a MERV rating of 8 or higher meets the Indoor Environmental Quality category's prerequisite for air filter quality. A higher MERV rating indicates a greater ability to remove particulate matter. However, a filter with a higher MERV rating may restrict airflow and reduce the speed of air filtration. A filter should be chosen based on the system's capacity and the project's requirements.

22. *The answers are:* **(A)** efficient appliances
(C) low-flow showerheads
(D) pipe insulation

Insulating hot water piping reduces the amount of heat lost as hot water travels from the source to the user. Thus, the heater does not need to take into account heat lost, and less energy is needed to heat the water. Installing efficient appliances such as Energy Star dishwashers and clothes washers reduces energy and water use. Installing low-flow showerheads and other fixtures controls the amount of hot water used and, therefore, heated. Although high-efficiency irrigation and rainwater harvesting systems conserve the use of potable water, they do not reduce the need for hot water or the energy required to generate it.

23. *The answers are:* (B) dishwashers
    (D) toilets
    (E) municipal wells

Graywater is untreated wastewater that does not contain any organic matter. Often it is collected from showers and baths, laundry machines, and bathroom sinks. For WE Credit 1.2, Water Reuse: Graywater Reuse System, the graywater must be reused for irrigation. Water from toilets and dishwashers is blackwater and cannot be reused. Municipal well water is potable and using it would not count toward meeting the requirements of this credit.

24. *The answers are:* (A) ACCA's *Manual J*
    (B) ASHRAE's *Handbook of Fundamentals*

EA Credit 2, Insulation, requires that insulation meet or exceed the *R*-value requirements of the ICC's International Energy Conservation Code, and meet the HERS Grade I and/or Grade II specifications. The insulation must then be verified by a green rater using the Energy Star Thermal Bypass Inspection Checklist. If a portion of a home is constructed with structural insulated panels (commonly called SIPs) or insulating concrete forms (commonly called ICFs), the green rater must use the Energy Star Structural Insulated Panel Visual Inspection Form.

25. *The answer is:* (B) The landscape design must include drought-tolerant plants.

SS Credit 2.2, Landscaping: Basic Landscaping Design, sets requirements for compacted soil and turf use, but does not specify plant selection. SS Credit 2.4, Landscaping: Drought-Tolerant Plants, requires installation of drought-tolerant plants.

26. *The answers are:* (A) the air tightness of the home's envelope
    (C) the duality of the home's insulation
    (D) the home's size

Envelope tightness, insulation, and home size are used to determine the appropriate size of HVAC equipment. Minimizing envelope leakage, improving insulation, and designing a smaller home will all help reduce a home's required HVAC system capacity. Neither a home's receptacle load (the maximum number of items that can be plugged into wall outlets), nor its finish materials (such as flooring) play a role in determining the appropriate size of a home's HVAC equipment.

27. *The answer is:* (C) Florida

Requirements for Energy Star windows are defined based on a home's location. Accordingly, the U.S. EPA has defined four climate zones for window requirements: Northern, North/Central, Southern, and South/Central. Since the LEED for Homes rating system's requirements for windows are based on Energy Star's requirements, the LEED for Homes rating system also uses these zones to set window requirements. EA Prerequisite 4.1, Good Windows, requires that homes in the South or South/Central climate zones with a window-to-floor area ratio of 18% or more meet a more stringent solar heat gain coefficient requirement than homes in the North and North/Central climate zones. Florida is in the South/Central climate zone, while Iowa, Pennsylvania, and Utah are in the North and North/Central climate zones.

28. The answer is: (B) U.S. EPA

The U.S. EPA's Energy Star Indoor Air Package recognizes homes with systems that ensure high standards of indoor air quality, and is referenced in EQ Credit 1, Energy Star with Indoor Air Package. The package sets requirements for multiple factors: moisture control, pest management, the HVAC system, combustion and venting systems, building materials, and radon control.

While ASHRAE is referenced and does set standards for indoor air quality, it does not have a package used by LEED for Homes. Green Seal's Green Label Plus and Greenguard Environmental Institute's certification program apply to products with low VOC emissions, but do not deal with indoor air quality specifically.

29. The answer is: (A) cork

Cork is a wood product that comes from the bark of the cork oak tree. While cork pieces are bound with resin to form many composite products (such as cork flooring), the raw material is not considered composite wood.

Composite wood includes a range of products manufactured by binding wood or plant particles, fibers, or veneers using a synthetic resin or adhesive. Plywood, oriented-strand board (commonly called OSB), and wheatboard are examples of composite wood. Though composite wood products may have environmental advantages over lumber, many are made with urea-formaldehyde resins, which have high VOC emissions.

30. The answers are: (C) 60% of hard-wired fixtures
(D) 100% of ceiling fans

EA Credit 8.3, Advanced Lighting Package, requires 60% of the hard-wired fixtures and 100% of the ceiling fans to be Energy Star-rated. Alternatively, a project can meet the credit requirements by installing Energy Star-labeled lamps in 80% of the home's fixtures.

31. The answer is: (C) graywater reuse system

Graywater recycling is not allowed in all jurisdictions. During project planning, a project team should consult state and local codes and apply for any necessary permits.

32. The answers are: (C) paints
(D) primers

LEED for Homes uses many standards to set VOC limits. MR Credit 2.2, Environmentally Preferable Products, uses Green Seal Standard GS-11 to set VOC limits for interior architectural paints, coatings, and primers applied to interior walls and ceilings. Green Seal Standard GS-36 sets VOC limits for commercial adhesives. SCAQMD's Rule 1113, sets VOC limits for stains sets. Carpets must meet the requirements of either the Green Label Plus program or the FloorScore program.

Practice Exam Part Two Answers

33. *The answers are:* **(D)** EA Credit 8, Lighting
 (E) EA Credit 9, Appliances

Installing Energy Star-rated ceiling fans can contribute the requirements of both EA Credit 8, Lighting, and EA Credit 9, Appliances. Ceiling fans are rated based on airflow and energy use at low and high speeds. Combination ceiling fan/light units are approximately 50% more efficient than conventional separate units.

34. *The answer is:* **(B)** HERS-trained energy raters only

EA Prerequisite 1.1, Optimize Energy Performance, requires the use of HERS energy modeling software, which is available only to HERS providers. To qualify for this credit, the software must be used only by HERS-trained energy raters. Green raters can be HERS trained energy raters, but they are not required to be.

35. *The answer is:* **(C)** an artificial wetland used for stormwater mitigation

Wetlands that are new and constructed for stormwater mitigation or site restoration are exempt from the requirements of LL Credit 2, Site Selection. Building on a natural wetland less than 10 years old, on a wetland within 100 feet of the home site, or on a wetland that is above the floodplain would make the home ineligible for this credit.

36. *The answers are:* **(B)** downspout
 (C) gutter

A rainwater harvesting conveyance system channels rainwater from a home's roof to storage compartments using gutters and downspouts. Tanks and cisterns store water, and pumps force the water through pipes and into an irrigation or other reuse system.

37. *The answers are:* **(B)** open space protected by local code
 (D) public parkland

Buildable land is defined as the portion of a site where construction can occur. It excludes public streets and other public rights-of-way, land occupied by nonresidential structures, public parks, and land excluded from residential development by law (such as that protected by local code). Though building on undeveloped land with native vegetation is not preferred, it is allowed. Projects cannot be constructed below FEMA's 100-year flood plain. Building above it is not restricted.

38. *The answer is:* **(A)** a material's ability to allow moisture to pass through it

A material's permeability is a measure of its ability to allow moisture to pass through it. Permeable surfaces (whether they are in the form of a vegetated landscape or permeable paving) help stormwater make its way back into the ground. For the purposes of LEED for Homes, permeable paving is made of porous above-ground materials over a 6 inch porous sub-base that ensures proper drainage away from a home. For SS Credit 4, Surface Water Management, 70% of the exposed site must be permeable.

39. *The answers are:* (A) Continuously operate the exhaust fan.

(C) Install an exhaust fan in the attic.

EQ Prerequisite 5.1, Basic Local Exhaust, requires meeting air flow requirements of ANSI/ASHRAE 62.2, installing bathroom and kitchen exhaust fans that are Energy Star-rated, and exhausting air to the outdoors. Continuous operation of an exhaust fan is part of EQ Credit 5.2, Enhanced Local Exhaust.

40. *The answer is:* (A) 0 points

Regardless of the climate zone, a home with a HERS rating of 95 is not eligible to earn points under EA Credit 1, Optimize Energy Performance. HERS ratings quantify a home's energy efficiency using an energy simulation model. Lower values represent higher efficiency. A HERS rating of 100 represents the energy efficiency of a home that meets basic International Energy Conservation Code (IECC) requirements. LEED for Homes EA Credit 1, Optimize Energy Performance, requires better energy efficiency than the IECC requirements. A home in IECC climate zones 1 through 5 can earn points for EA Credit 1 by having a HERS rating of 84 or lower. A home in IECC climate zones 6 through 8 can earn points for EA Credit 1 by having a HERS rating of 79 or lower.

41. *The answers are:* (B) preparing detailed framing plans

(D) using off-site fabrication

MR Credit 1, Material-Efficient Framing, lists acceptable strategies for efficient framing. Detailed framing documents help a project team accurately determine the required size and number of framing members, which results in less wasted material. Fabricating panelized, modular, or prefabricated components off site can also reduce waste. Recycling excess material and using FCS-certified or salvaged wood are sustainable strategies, but do not directly contribute to material-efficient framing requirements.

42. *The answers are:* (A) carbon monoxide monitors on each floor

(B) doors on all fireplaces

EQ Credit 2, Combustion Venting, lists combustion venting measures that will reduce the leakage of combustible gases into a home. Recommended basic measures include installing doors on all fireplaces and carbon dioxide monitors on each floor of the home. Not installing fireplaces is an enhanced (not basic) venting measure, and unvented combustion appliances are not permissible. Installing wood burning stoves is not a combustion venting measure, but if they are installed, the installation must meet specific credit requirements.

43. *The answer is:* (B) hydronic system

A hydronic system is a non-ducted HVAC system that uses circulating water for cooling or heating purposes. To meet EA Prerequisite 5.1, Heating and Cooling Distribution: Reduced Distribution Losses, a hydronic system's distribution pipes in unconditioned areas must be insulated with R-3 insulation or better. Forced air systems require R-6 insulation or better around ducts in unconditioned spaces. All domestic hot water piping, as well as the central manifold system, must have R-4 insulation or better.

44. *The answers are:* **(A)** increased mass transit use
(B) increased pedestrian activity
(D) protection of undeveloped land

Compact land development promotes community connectivity, which increases mass-transit use and pedestrian activity. It also conserves and protects undeveloped land. Compact land development does not specifically protect endangered species or reduce the amount of water used for irrigation in the community landscape.

45. *The answer is:* **(B)** 5000 gallons

WE Credit 1.2, Water Reuse: Graywater Reuse System, requires a graywater system to collect at least 5000 gallons of water per year from clothes washers, showers, and a combination of other applicable faucets.

46. *The answer is:* **(C)** Carpet and Rug Institute

The Carpet and Rug Institute manages the Green Label and Green Label Plus programs, which tests the VOC emissions of carpets, cushions, and adhesives, and identifies products with low and very low VOC emissions.

47. *The answer is:* **(A)** 1.0 cu ft

EA Credit 5.3, Heating and Cooling Distribution: Minimal Distribution Losses, requires that the duct leakage rate be tested. It can be no more than 1.0 cu ft per minute at 25 Pa per 100 sq ft of conditioned floor area and must be verified by an energy rater. Other acceptable approaches for meeting the credit requirements include installing the air handler and all ductwork within the conditioned envelope and minimizing leakage, or installing the air handler and all ductwork in sight (not hidden in walls, floors, etc.) and within conditioned spaces.

48. *The answers are:* **(B)** constructed wetland
(C) pond
(D) rain garden

WE Credit 1.1, Water Reuse: Rainwater Harvesting System, requires that rainwater be harvested in an opaque cistern or tank out of the sun to minimize algae, bacteria growth, pests, and debris. If implemented effectively, rainwater harvesting systems can reduce runoff and erosion and replace potable water used for landscape irrigation and some indoor water use. Constructed wetlands, ponds, and rain gardens can also capture rainwater and prevent runoff and erosion, but water from them cannot be reused for the same purposes as properly harvested and stored rainwater.

49. *The answer is:* **(B)** irrigation systems

The U.S. EPA's WaterSense program certifies professionals to design and install irrigation systems. While the program does not certify installers of high-efficiency fixtures and fittings, it does certify the fixtures and fittings themselves.

50. *The answers are:* (A) It is a byproduct of burning coal.

(B) It is sometimes a component of concrete.

Fly ash is a byproduct of burning coal and can be recycled and used in making concrete, which can help a project earn points for MR Credit 2, Environmentally Preferable Products.

51. *The answer is:* (C) sizing duct systems

Understanding airflow and proper duct design is an essential part of designing an appropriate HVAC system. ACCA's *Manual D*, the nationally-recognized standard for designing residential HVAC duct systems, is referenced in EQ Prerequisite 6.1, Distribution of Space Heating and Cooling: Room-by-Room Load Calculations.

52. *The answer is:* (A) a home's air conditioning system

A seasonal energy efficiency rating (SEER) measures air conditioner and heat pump efficiency. It is calculated using the amount of cooling (or heating) output, measured in British Thermal Units (BTUs) during a typical cooling (or heating) season, divided by the energy usage, measured in watt-hours during the same season. The higher the SEER, the greater the efficiency and, therefore, the greater the energy savings. Since 2006, U.S. regulatory agencies have required new air conditioners to have a SEER of 13.0 or better.

HERS rates the energy efficiency of an entire home, MERV rates the efficiency of air filters, and Energy Star rates the energy efficiency of lighting fixtures.

53. *The answer is:* (A) Forest Stewardship Council

The Forest Stewardship Council (FSC) offers chain-of-custody certification, which indicates that a certified product has met FSC guidelines from point of extraction to point of sale.

54. *The answer is:* (B) surface reflectivity

Albedo is a measure of a surface's reflective ability; it is the fraction of incident radiation (light) that a surface reflects. Albedo values range from 0 (black) to 1 (white). A material with a high albedo reflects the majority of light that hits it. For the purposes of LEED, surfaces such as roofs and hardscapes with a high albedo are preferable. Such surfaces do not absorb heat as readily and thus do not create heat islands, which can increase a home's energy demands for space cooling.

55. *The answers are:* (A) demolition debris

(B) grass clippings

(C) material collected through recycling programs

Post-consumer waste is material generated by households or commercial, industrial, or institutional facilities, and that can no longer be used for its original purpose. Demolition debris, grass clippings, and material collected through recycling programs are examples of post-consumer waste. Sawdust and metal trimmings are considered pre-consumer content since they are diverted from the waste stream during the manufacturing process.

56. *The answers are:* (A) drought-tolerant plants
(C) native plants
(F) indigenous plants

Native plants, also called indigenous plants, are those originating from a given geographical region. Drought-tolerant plants need a relatively small amount of water, thus minimizing the need for supplemental irrigation. Xeriscaping is a method of landscaping that involves preparing the soil to reduce water loss to evaporation and runoff, eliminating the need for supplemental irrigation, and planting appropriate species. The specific plant species used in xeriscaping varies depending on the local climate, but xeriscaping generally includes the use of native, indigenous, and/or drought-tolerant plants.

57. *The answer is:* (C) 8 climate zones

According to the International Energy Conservation Code, there are 8 climate zones in the U.S. The 8 climate zone classifications are primarily defined based on temperature and humidity levels. Climate zone 1 is the hottest and climate zone 8 the coldest.

58. *The answer is:* (B) calculating heating and cooling loads

EA Credit 6, Space Heating and Cooling Equipment, requires design calculations to be performed using ACCA's *Manual J* and *Manual D*, ASHRAE's *Handbook of Fundamentals*, or an equivalent. *Manual J: Residential Load Calculation* provides guidelines for calculating heating and cooling loads and for sizing HVAC equipment.

59. *The answer is:* (B) 2 points

A project can earn points for reducing appliance energy consumption by complying with the requirements of EA Credit 9, Appliances. The project will earn 1 point for installing one or more Energy Star refrigerators, a half point for installing an Energy Star clothes washer, and a half point for installing an Energy Star dishwasher, for a total of 2 points.

60. *The answer is:* (D) vegetated swale

A rain garden is a constructed low tract of land that is vegetated, collects rain water, and tolerates high moisture levels. A vegetated swale is similar, but may be either constructed or naturally-occurring. When it is constructed, its intended purpose is the same as that of a rain garden. Both rain gardens and vegetated swales prevent stormwater runoff and replenish groundwater.

61. *The answers are:* (B) graywater
(C) harvested rainwater
(F) reclaimed water

Graywater, reclaimed wastewater, and harvested rainwater can be used for below-ground irrigation, but are not recommended for above-ground irrigation, and should never be used for watering edible plants, such as those in a vegetable garden, because of the risk of contamination. Harvested rainwater is runoff that is collected in a cistern. Graywater is untreated wastewater that does not contain any organic matter. Often it is collected from showers and baths, laundry machines, and bathroom sinks. Reclaimed wastewater, sometimes called irrigation quality water, is another possible source for irrigation water. Reclaimed water is processed in a wastewater treatment plant, but is only minimally treated to make it suitable for landscape irrigation and other nonpotable uses such as cooling towers, industrial process uses, toilet flushing, and fire protection.

Blackwater is wastewater contaminated with organic matter and cannot be reused; this includes water from toilets, urinals, kitchen sinks, dishwashers, and sometimes washing machines. Using municipal or potable water for irrigation purposes is only considered sustainable if other water use reduction strategies are implemented simultaneously.

62. The answer is: (C) 60%

EQ Credit 3, Moisture Control, requires that the relative humidity of a home be maintained at or below 60%. If this is not possible naturally, dehumidification equipment or a central HVAC system equipped to operate in dehumidification mode should be installed.

63. The answer is: (B) 1400 sq ft

The LEED for Homes rating system uses a home size adjustment calculation to establish certification thresholds. This is done with the assumption that a larger home consumes more resources than a smaller home (all other things being equal). An "average-sized" home is considered neutral. The point thresholds for each certification level vary depending on whether a home's size is greater than, equal to, or less than the size of a neutral home. The following are neutral home sizes.

    1 bedroom: 900 sq ft

    2 bedrooms: 1400 sq ft

    3 bedrooms: 1900 sq ft

    4 bedrooms: 2600 sq ft

    5 bedrooms: 2850 sq ft

For homes with more than 5 bedrooms, 250 sq ft is added for each additional bedroom.

A larger home must earn more points to reach each certification level threshold, and a smaller home needs fewer points to reach each threshold. The following table demonstrates the varying point thresholds for three sizes of a 1 bedroom home.

1 bedroom homesize	certification thresholds			
	Certified	Silver	Gold	Platinum
800 sq ft	42–56 points	57–71 points	72–86 points	87–133 points
900 sq ft (neutral)	45–59 points	60–74 points	75–89 points	90–136 points
1010 sq ft	48–62 points	63–77 points	78–92 points	93–139 points

64. The answer is: (A) back-draft potential

An Energy Star home must pass a visual inspection of energy saving measures, an envelope leakage performance test, and a thermal bypass inspection. The mandatory performance tests include envelope and duct tightness but not back-draft potential.

65. *The answers are:* **(A)** biofuel systems
(B) geothermal systems
(D) photovoltaic systems

Renewable energy is generated from naturally replenishing resources such as sun, wind, rain, tides, biofuels, and geothermal heat. Photovoltaic systems use the sun to generate energy, and geothermal systems extract energy from heat stored in the earth. Though passive solar heat systems and wood furnaces use renewable energy sources, they do not generate energy.

66. *The answers are:* **(B)** if the land has deeded public access
(D) if there is a precedent of public use and 10 year commitment to future public use

Private land can be considered public open space and can be used for passive recreation as long as there is deeded public access or a history of public use that is anticipated to continue for 10 years. Within the LEED for Homes rating system, public open space is addressed in LL Credit 6, Access to Open Space.

67. *The answer is:* **(C)** U.S. EPA

The U.S. Environmental Protection Agency (U.S. EPA) sets the Energy Star program's energy efficiency guidelines, which USGBC refers to in several LEED for Homes credits refers to as a benchmark for energy savings.

68. *The answer is:* **(D)** Turning on all dehumidification equipment during the preoccupancy flush

A preoccupancy flush is conducted prior to occupancy but after all construction is complete. The home must be flushed with fresh air according to requirements of EQ Credit 8.2, Contaminant Control: Indoor Contaminant Control. The flush must last 48 consecutive hours with all interior doors open, windows open, and fans running. Air filters must be cleaned or replaced after flush. Dehumidification equipment should not be turned on during the flush.

69. *The answer is:* **(D)** 60 sq ft

EA Credit 4.1, Good Windows, requires that the ratio of skylight glazing to the conditioned floor area not exceed 3%. For a home with 2000 sq ft of conditioned floor area, this means the maximum area of skylight glazing is 60 sq ft.

70. *The answer is:* **(A)** within 500 miles of the home

A home can earn a point for MR Credit 2.2, Environmentally Preferable Products, for using local or regional materials, which the LEED for Homes reference guide defines as those that are extracted, processed, and manufactured within 500 miles of the home.

71. *The answer is:* **(B)** fundamental commissioning

The LEED for Homes program requires green rater inspections as an alternative to the commissioning needed for other LEED certification programs; therefore there is no prerequisite for commissioning in the LEED for Homes rating system. Durability planning is a prerequisite in the Innovation and Design credit category; the framing waste factor must be reduced by at least 10% to comply with MR Prerequisite 1.1, Material-Efficient Framing: Framing Order Waste Factor.

72. *The answer is:* **(C)** house footprint

SS Credit 4.1, Surface Water Management: Permeable Lot, requires documentation of the area of a home site's built environment. This includes walkways, driveways, designed landscape, and undisturbed buildable lot locations. The space under the roof of the home, garage, or any other structure, and public rights-of-way and roads are excluded.

73. *The answer is:* **(A)** edge development

To comply with the requirements of the LL Credit 3.1, Preferred Locations: Edge Development, at least 25% of an edge development's perimeter must border an established community.

74. *The answers are:* **(C)** Nicaragua
(D) Philippines

Wood grown in a location south of the Tropic of Cancer and north of the Tropic of Capricorn complies with the location requirements for MR Prerequisite 2.1, Environmentally Preferable Products: FCS Certified Tropical Wood. Central America (including Nicaragua) and the Philippines are within these limits. There are no states within the U.S. that fall within these limits. Spain is north of the Tropic of Cancer and South Africa is south of the Tropic of Capricorn.

75. *The answer is:* **(A)** a five-story multifamily condominium

Project types eligible for LEED certification under the LEED for Homes rating system include single-family homes, low-rise (one- to three-story) multifamily buildings, production homes, affordable homes, manufactured and modular homes, and existing homes.

76. *The answer is:* **(D)** the number of dwelling units per acre of buildable land

Within SS Credit 6, Compact Development, the LEED for Homes reference guide defines density for residential areas as the number of dwelling units per acre of buildable land available for residential uses. For nonresidential areas, the reference guide defines density as the floor area per net acre of buildable land available for nonresidential uses.

77. *The answer is:* **(B)** minimizing east-west sun exposure

A home that is oriented to minimize sun exposure on east- and west-facing walls will require relatively less mechanical heating and cooling, and thus this strategy will reduce the need for energy. The decision to minimize east-west sun exposure must be made during the site analysis, because implementing the decision affects the way that the foundation of the home is laid. While maximizing insulation and minimizing air leakage and thermal bridges should be planned early, the foundation of the home can be laid beforehand.

78. *The answers are:* (C) reclaimed material
    (E) reused material
    (F) salvaged material

USGBC defines reclaimed, reused, and salvaged material as any building component that has been recovered from a demolition site and is reused in its original state, thereby diverting it from the waste stream. Waste (such as construction waste) is material that is not diverted from the waste stream. Preconsumer material is diverted from the waste stream during the manufacturing process. It is generally converted from its original post-manufacturing state into something consumable (e.g., wood chips from a lumber mill used in particle board).

79. *The answer is:* (A) albedo

Albedo and solar reflectance index (SRI) are measures of a surface's reflectivity. A material with a high SRI value will also have a high albedo value. Using high-albedo materials will contribute toward SS Credit 3, Local Heat Island Effects.

80. *The answer is:* (D) It causes erosion.

Sedimentation is the deposition and accumulation of soil and other natural solids in water bodies, and it is caused by stormwater runoff and erosion. It accelerates the rate at which lakes, streams, and rivers fill in, called aging. It is similar to but not the same as siltation, which is the accumulation of fine particles in water bodies.

81. *The answers are:* (A) three-stud corners
    (B) floor joist spacing at 16 inches on center

MR Credit 1.4, Material-Efficient Framing: Framing Efficiencies, lists appropriate strategies to reduce the waste factor of framing. Some efficient strategies include installing corners with two studs (not three) and spacing floor joists at more than 16 inches on center. Project teams must conform to local building codes and regulations, implementing only those efficient framing measures that are permissible. If open web truss joists and structural insulated panels are fabricated off site, this also reduces the framing waste factor.

82. *The answers are:* (A) appliance selection
    (C) landscaping
    (D) lighting selection

AE Credit 2, Education of Building Manager, requires the building manager of a residential dwellings with more than five units to be trained to maintain the building's overall performance. The manager must also provide guidance to occupants to reduce water use and save energy, which requires advice on appliance selection, landscaping, and lighting selection. AE Credit 2 does not require the building manager to advise occupants on furniture selection, interior decoration, or the size of unit to rent or purchase.

83. The answer is: (C) if the site has been previously developed

LL Credit 2, Site Selection, discourages development on land with unique soils, prime soils, or soils of state significance. A home built on these types of land can only earn points for this credit when the selected site has been previously developed. A soil of state significance is a a soil that the National Resources Conservation Service has specified as having special value for a particular state.

84. The answers are: (A) minimizing the area of impermeable hardscape
(C) installing a vegetated roof
(E) installing permeable walkways

Stormwater retention can be improved by increasing the permeable area of a lot. Strategies to do so include minimizing the area of impermeable hardscape, installing permeable paving, installing a vegetated roof, and reducing the under-roof area. Installing drought-tolerant plants can reduce the need for irrigation but does not help manage stormwater runoff.

85. The answer is: (B) a table made from cedar felled by a storm

Reclaimed materials are building components recovered from a demolition site and reused in their original state. In the LEED for Homes rating system, reclaimed material is construction site debris that has previously been used by a consumer (i.e., not the scraps from construction) and is diverted from the waste stream. Cedar felled by a storm is preconsumer material, and thus would not be considered reclaimed.

86. The answer is: (D) 38 points

The maximum number of points that a project may earn in the Energy and Atmosphere credit category is 38.

87. The answer is: (C) reduced home construction costs

A Planned Unit Development (PUD) is a floating overlay district that meets overall community density and land use goals without being bound by existing zoning requirements. PUDs generally have common open space, include a variety of building types and land uses, and use building clustering. Benefits of a PUD often include more efficient site design, preservation of open space, and lower costs for street construction, utility extension, and maintenance.

88. The answer is: (B) 1.3 gallons of water per flush

High-efficiency toilets use no more than 1.3 gallons of water per flush. High-efficiency urinals use no more than 1 gallon of water per flush.

89. The answer is: (B) in a faucet

Aerators add air to water to increase the perceived amount of water flowing through the faucet.

Practice Exam Part Two Answers

90. *The answers are:* **(B)** EA Credit 7.2, Water Heating: Pipe Insulation

    (F) EA Credit 11.2, Residential Refrigerant Management: Appropriate HVAC Refrigerants

When a project earns points for EA Credit 1, opportunities for earning additional Energy and Atmosphere points are limited to: EA Credit 7.1, Water Heating: Efficient Hot Water Distribution; EA Credit 7.2, Water Heating: Pipe Insulation; and EA Credit 11.2, Residential Refrigerant Management: Appropriate HVAC Refrigerants. The only Energy and Atmosphere prerequisite for projects pursuing EA Credit 1, Optimize Energy Performance is EA Prerequisite 11.1, Residential Refrigerant Management: Refrigerant Charge Test.

91. *The answers are:* **(A)** fountains

    (B) gravel

    (C) paving stones

Hardscaping consists of non-vegetative surfaces and human-made elements usually surrounded by landscaping. Examples of hardscape elements include paving materials, roads, yard sculptures, fountains, and retaining walls.

92. *The answer is:* **(D)** REScheck

EA Prerequisite 2.1, Basic Insulation, requires that the insulation level of a home meet or exceed *R*-value requirements in Chapter 4 of the International Energy Conservation Code (IECC). REScheck clarifies the Model Energy Code, the IECC, and a number of state codes.

93. *The answer is:* **(B)** evapotranspiration factor

The landscape coefficient is the result of multiplying the species factor, density factor, and microclimate factor of a particular location. The landscape coefficient is then used to calculate the evapotranspiration rate for WE Credit 2, Irrigation System.

94. *The answers are:* **(C)** to promote mass transit use

    (D) to promote pedestrian activity

    (E) to protect undeveloped land

Complying with the requirements of SS Credit 6, Compact Development, promotes walkability, transportation efficiency, and community connectivity, and protects undeveloped land from development. Compact development will not necessarily protect endangered species or reduce irrigation demand.

95. *The answer is:* **(C)** 84

Point earning in EA Credit 1, Optimize Energy Performance, is determined using the International Energy Conservation Code (IECC) climate zones, IECC energy performance levels, and HERS indices. A home in IECC climate zones 1–5 with a HERS index of 84 or less (or a home that performs 16% better than the IECC requirements) qualifies to earn points for EA Credit 1. A home in IECC climate zones 6–8 with a HERS index of 80 or less (or a home that performs 21% better than the IECC requirements) qualifies to earn points for EA Credit 1.

96. *The answer is:* **(D)** R410a

EA Credit 11.2, Residential Refrigerant Management: Appropriate HVAC Refrigerants, allows the use of some refrigerants that do not contain chlorofluorocarbons (CFCs) or hydrochlorofluorocarbons (HCFCs), including R410a. USGBC prohibits the use of CFCs because they are proven to be a major cause of depletion of the earth's stratospheric ozone layer and contribute to the greenhouse effect and global warming.

97. *The answers are:* **(B)** hard ducts on all returns
(D) sealed duct seams

EA Credit 5, Heating and Cooling Distribution System, sets requirements for implementation of forced air systems and non-ducted systems. In a forced air system, the duct leakage rate must be tested and verified by an energy rater, there must be hard ducts on all returns, and the duct seams and joints must be sealed with water-based mastic. Ducts in unconditioned spaces must also have R-6 insulation or better.

98. *The answers are:* **(A)** increasing the shaded area
(C) installing white concrete
(D) installing open pavers

To comply with the requirements of SS Credit 3, Local Heat Island Effects, trees and other plantings must provide shade for at least 50% of the hardscape within 50 feet of the home. Alternatively, the project can use high-albedo (light colored) materials, open pavers, or vegetation for at least 50% of the hardscape within 50 feet of the home.

Installing drought-tolerant plants and limiting the use of conventional turf will contribute to SS Credit 2, Landscaping, and WE Credit 2, Irrigation (but will not help reduce the heat island effect). The result of reducing light pollution is reducing sky glow at night. Though this benefits the nocturnal environment, it will not help reduce heat islands, nor is it a LEED for Homes point-earning strategy.

99. *The answers are:* **(B)** branch line diameter
(E) trunk length

EA Credit 7.1, Efficient Hot Water Distribution, lists a central manifold distribution system as a hot-water distribution option as long as the system meets specific criteria. The central manifold trunk must have at least R-4 insulation and must be no more than 6 inches long. The branch lines must have a maximum nominal diameter of 0.5 inches and may not exceed 20 feet to any fixture in a one-story home.

100. *The answer is:* **(B)** choosing a previously developed lot

Choosing a previously developed lot will earn a project a point for LL Credit 3, Preferred Locations, but it will not earn the project points within the Sustainable Sites category. Controlling pests is required for SS Credit 5, Nontoxic Pest Control. Choosing a compact lot is required for SS Credit 6, Compact Lot Development. Reducing irrigation demand can earn a project points in both the Water Efficiency and the Sustainable Sites categories.